Discover the Moon

Jean Lacroux
Christian Legrand

CAMBRIDGE
UNIVERSITY PRESS

PUBLISHED BY THE PRESS SYNDICATE OF THE UNIVERSITY OF CAMBRIDGE
The Pitt Building, Trumpington Street, Cambridge, United Kingdom

CAMBRIDGE UNIVERSITY PRESS
The Edinburgh Building, Cambridge CB2 2RU, UK
40 West 20th Street, New York, NY 10011–4211, USA
477 Williamstown Road, Port Melbourne, VIC 3207, Australia
Ruiz de Alarcón 13, 28014 Madrid, Spain
Dock House, The Waterfront, Cape Town 8001, South Africa

http://www.cambridge.org

Originally published as *Découvrir la Lune*, by J. Lacroux and C. Legrand
© Bordas/VUEF 2000
This translation © Cambridge University Press 2003
Published with the help of the French Ministry of Culture

First published 2003

Printed in the United Kingdom at the University Press, Cambridge

Typeface Berthold Garamond 10.5/12   System QuarkXPress™   [SE]

A catalogue record for this book is available from the British Library

Library of Congress Cataloguing in Publication data

Lacroux, Jean.
  [Découvrir la lune. English]
  Discover the moon / Jean Lacroux and Christian Legrand.
    p.   cm.
  Includes bibliographical references and index.
  ISBN 0 521 53555 7 (pbk.)
  1. Moon – Observers' manuals.   I. Legrand, Christian.   II. Title.

  QB581.L3313   2003
  523.3–dc21   2002041534

ISBN 0 521 53555 7 paperback

The publisher has used its best endeavours to ensure that the URLs
for external websites referred to in this book are correct and active at
the time of going to press. However, the publisher has no
responsibility for the websites and can make no guarantee that a site
will remain live or that the content is or will remain appropriate.

Preface

Ever looked at the Moon through a telescope? You have? Then you will have felt 'astronomical awe' for yourself.

The Moon . . . It is the strangest place! A rough, dry mineral sphere with a cloudless sky that is inky black even in bright daylight, waterless seas decked in dust that no winds ever blow, and worn mountains that have never echoed to the slightest sound . . . It is understandable, then, why our satellite should be the favourite target for aspiring astronomers. And it is to help them to become better observers and to enjoy their discoveries to the full that this book has been devised.

It is all very well to stand and stare, but it is so much better to understand what you are looking at. You will want to see the most interesting and most intriguing regions of the Moon. But how do you find them in your telescope's field of view? When is the best time to look for them?

Then, with a little experience, you will be able to keep a watch on places where 'something' might be going on . . .

The Moon is easy enough to observe even with the light pollution of modern cities. Even the smallest telescope will show the maria or 'seas', countless craters and a few mountain ranges. The Moon's spectacular relief and the wondrous calm of its desolate landscapes viewed through a telescope lend it a fascination you will never tire of.

In this book we do not try to present an exhaustive survey of lunar studies; we try simply to answer some of the questions of the 'moonstruck' by providing material to assist in observation. This explorer's guidebook will help them to find their way around.

So, to your telescopes, for some fantastic trips to the Moon!

Contents

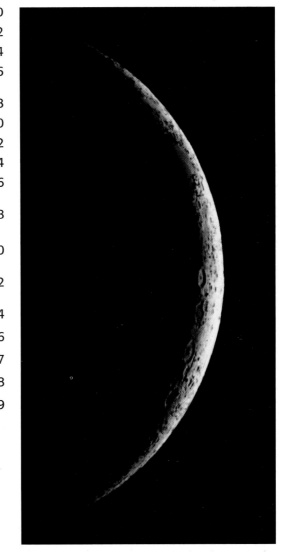

How to use this book

*T*his guidebook is devised to make it easier to identify and observe the most interesting lunar features. It uses two sets of photos showing the Moon as it appears through the three types of instrument most commonly used by amateur astronomers – refracting telescopes, catadioptric telescopes and Newtonian telescopes. This has not been done before.

Which way is up?

It is often difficult to use a map to locate a lunar feature in a telescope. The thing is, lunar charts show the Moon as it appears to the naked eye (and have done since 1961; see box). If you view the Moon through binoculars there is no difference with the map because binoculars do not invert the image but, instead, show the Moon just as it looks to the unaided eye.

But when the Moon is viewed through any astronomical telescope (be it a refractor or a reflector) without accessories, the images are not the same way round. The objective lens or mirror produces an inverted image.

In a reflecting telescope, the secondary mirror inside it alters the image again! So, with a Newtonian telescope like the famous 115-mm (4.5-inch) model, the image is completely inverted with north at the bottom and east on the left. A lunar map has to be turned through 180° if it is to show the same alignment.

However, with instruments that have a star diagonal such as astronomical refractors or Cassegrain, Maksutov or Schmidt–Cassegrain catadioptric telescopes (like the Celestron or Meade makes), the image is erected by the star diagonal so that the north of the Moon is at the top and south at the bottom, but east is still on the left and west on the right! This time you would need to view the map in a mirror held alongside it for the chart to match what you see through this type of telescope!

DON'T PANIC!

It may be that what you observe on any given night does not exactly match the photo in the book. Some of the craters may not be illuminated in quite the same way. This is because of librations that make the Moon 'rock and roll' (see p. 15).

The photos are there to help you locate the positions of features relative to each other. The appearance of any given landform will always be different with each phase of the Moon. This is what makes discovering the lunar surface so fascinating time and again.

A SENSE OF DIRECTION

In 1961, the International Astronomical Union stipulated that lunar maps should have north at the top, south at the bottom, east on the right and west on the left. But do not be surprised if you find that older books do the opposite. So now you see why Mare Orientale (the Eastern Sea) has found itself on the Moon's western edge!

As seen through the telescope

This book is unique in that it overcomes these difficulties of orientation by presenting two photos for each lunar region or site. The photos are actually the same but oriented differently:

 • the left-hand page shows the view through a refracting telescope or a catadioptric telescope with a star diagonal

 • the right-hand page shows the view through a Newtonian telescope.

Night-by-night, 14 guided observing sessions

• A guidebook based on the Moon's phases

Each chapter presents the Moon on a different night throughout the series of phases from New Moon to Full Moon. We have given precedence to evening observations. Most observers prefer this because getting up early is harder than staying up late!

For each evening's observation there is a general photo first, with both possible orientations depending on the telescope used. The locations of features of interest to observe that night are marked by numbers. Boxed regions are described in more detail in the pages that follow.

• Detailed descriptions of characteristic regions

Each of the following double-page spreads in the chapter is about a particular region. The important features are precisely described so you know what to look for.

The photos show details down to distances of about 3 km. This way you can test the resolution of your telescope.

• Feature size indications

Each detailed photo has a scale bar so you can compare the dimensions of the different features and get some idea of the true size of these lunar landscapes.

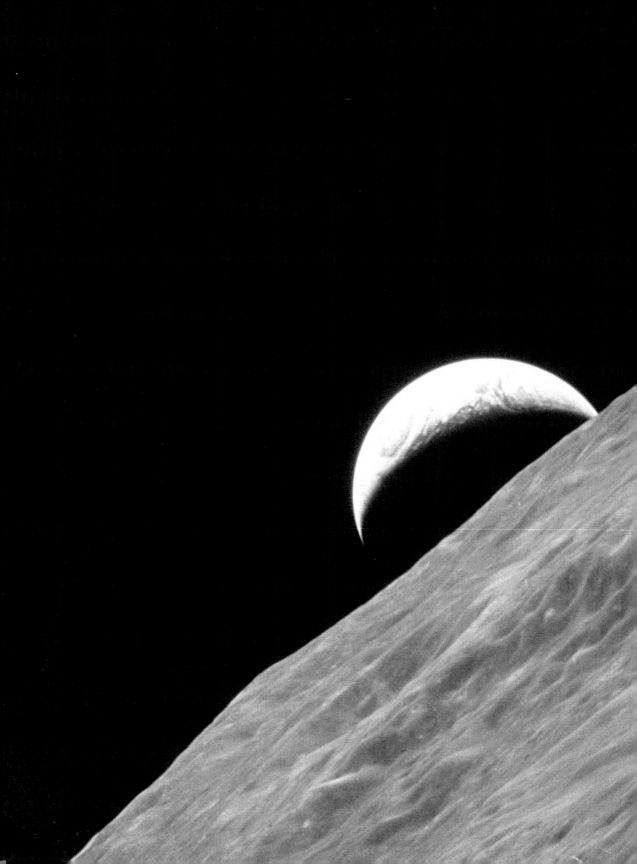

Earth's Moon

What is there to see on the Moon?

*E*ven the unaided eye gives a hint of what the lunar surface is like. Large dark patches can be made out, which early observers called 'seas', with brighter areas, known as 'highlands', between them. The practised eye can even make out a few bright spots within the seas.

Waterless seas

The Moon's surface is characterised by its 'seas' or maria, so named by the early astronomers who saw them as the counterparts of the Earth's oceans. We now know, though, that the maria contain no water in liquid form. They are vast, mostly flat expanses of basalt some 3.8 to 3.1 billion years old. They cover 17% of the lunar surface and very many more of them lie on the near side than on the far side of the Moon.

These maria are probably the result of giant meteorites striking the Moon some 600 million years after it first began to form.

The lunar maria are pitted with numerous craterlets like this one.

3 1833 04697 2631

WHERE DOES THE MOON COME FROM?

There are a number of theories about how our satellite originated:
- Roche's 1873 'double-planet hypothesis' by which the Moon accreted from the same cloud of dust as the Earth.
- The 'fission hypothesis' that it formed by a bulge of soft material spinning off from the still-molten primitive Earth as proposed by George Darwin in the 1880s.
- The 'capture hypothesis'. A suggestion first made by Lee in 1909 is that the Moon formed beyond the orbit of Uranus, moved closer to Earth because it was slowed by dust particles cluttering the solar system and was finally captured.
- The 'giant impact hypothesis' according to which it was torn off the Earth as suggested by Hartmann and Davis in 1974. This theory is based on the composition of lunar rocks brought back by the Apollo and Luna missions because they contain terrestrial elements and 'extraneous' elements. It is thought that the Moon was produced by a collision between the newly formed Earth and a mini-planet in formation some 4 billion years ago. The impact supposedly tore debris from the Earth which mixed with the material of the planetoid to form the Moon.

It is thought that these meteorite strikes pierced the primitive crust causing the still-molten rocky mantle to spill out over the surface.

Marine ridges

The maria are made up of basaltic lava with high proportions of iron, titanium and magnesium. Their surfaces are often ridged by very elongate, low hills known as 'dorsa' or 'wrinkle ridges', which sometimes branch out. Although only a hundred or so metres high, wrinkle ridges may extend for several thousand kilometres. They are thought to have formed by compression of the terrain as the surfaces of the maria cooled.

Gently sloping mountains

The Moon has many mountains that are the remains of the primitive crust. There are also entire mountain ranges which are the rims of the impact basins where the maria formed. Lunar mountains slope gently at gradients of 15–20°, very occasionally reaching 30–35°.

Some 20 or so mountain ranges have been catalogued.

There are also isolated mountains, generally occurring as peaks emerging from the lava of the maria. They are remnants of the initial underlying surface before it was covered by the molten basalt. About 15 isolated mountains are recorded.

Countless craters

The most characteristic features of the lunar surface are the countless meteorite craters ranging in size from 300 km down to less than 1 m in diameter. The near side of the Moon has more than 300 000 craters of more than 1 km in diameter.

A distinction is made by size between walled plains, classical craters and craterlets, but it is sometimes difficult to know quite which category to classify a formation in.
• Walled plains are large, often dilapidated and deformed mountainous rings of anything from 100 to 300 km in diameter. Their floors, which are often flat with many craters, craterlets, ridges and hills, sometimes follow the curvature of the Moon.
• Craters proper range from 10 to 100 km across. They have three separate parts: the outer slopes, the inner wall and the floor. The outer slopes are made up of ejecta and rise gently from the surrounding terrain to the often steep crater rim. The inner wall has ledge-like terraces in craters wider than 50 km. The floor is often flat

The Apollo missions confirmed that the lunar mountains were smooth and rounded.

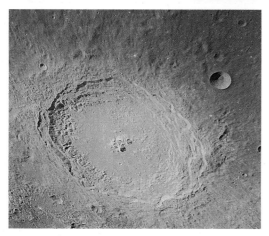

Close-up of a large crater clearly showing the gently dipping outer slopes, the slumped terraces of the crater wall and its rather rugged floor with a central peak.

with one or more central peaks or sometimes an inner ring of mountains. It is commonly cluttered by material or may be scarred by narrow, branching rilles.
• Craterlets are very prominent circular formations no more than 10 to 20 km across, with bowl-shaped floors.

Crater chains

Although rare, a few crater chains are found on the Moon. It is highly likely that all the craters in a chain were formed by the same event since there is a very low probability of a dozen or so craters forming a regular alignment over the course of time. Crater chains are thought to result from impacts from a single meteorite that broke up into a number of pieces just before striking the Moon.

Enigmatic clefts

Rilles are another typical formation. They are sometimes sinuous, branching furrows running for several hundred kilometres. Comparisons with the few specimens found on Earth suggest that they are ancient underground tunnels that once conveyed lava but whose roofs have since collapsed.

Other straighter clefts are grabens formed where plates of the lunar crust have moved apart.

Spectacular scarps

The lunar surface counts a handful of magnificent tectonic faults that show up marvellously when illuminated at a low angle when near the terminator. They rise to a few hundred metres and may be more than 100 km long. They are never sheer; indeed, most of them dip at less than 45°.

DON'T FORGET YOUR SPACESUIT!

The Moon has only a pseudo-atmosphere composed of traces of helium from the degassing of rocks and from their erosion by the solar wind. Atmospheric pressure is not even one-millionth of the Earth's. A number of phenomena support this hypothesis: the edge of the Moon appears very sharply with no blurring; the terminator has no twilight zone; no clouds hide the Moon's surface; stars vanish instantly behind the Moon's disc when occulted. This absence of any atmosphere has a number of consequences: there is no water, no wind and no noise; the sky is not blue but black, with the Sun shining beside the coloured stars and the blue-tinted Earth; temperatures are extreme (from +100 °C in daytime to −150 °C at night); there are no shooting stars but meteorite falls; there is a constant rain of micro-meteorites (several tens of tonnes of dust per day).

↑ *This flat-floored lunar rille was probably cut as the expanse of lava it crosses cooled. It is no more than 1 km deep and averages 5 km in width.*

← *Inside crater Davy Y, a rare chain of craterlets ranging from 1 to 3 km in diameter.*

Riverless valleys

The Moon's valleys were not cut by rivers or glaciers. They are generally rough alignments of craters of various sizes that overlap forming an elongated depression with an uneven floor. However, one lunar valley – the Alpine Valley – is a true graben, while Schröter's Valley is an exceptionally long rille.

Inconspicuous domes

Domes are rounded hills, sometimes with a crater at their summit. They are some 10 km in diameter but no more than 1 km high. It is thought that the domes with summit pits are extinct volcanoes and that the craterless ones are volcanic shields. No lava flows from the domes are apparent. They were probably formed by viscous magma expanding beneath the lunar crust long ago.

MOON LOVERS

A number of astronomy associations have special sections for lunar observation. The best known are the Association of Lunar and Planetary Observers (ALPO http://www.lpl.arizona.edu/~rhill/alpo/alpo_index.html), the British Astronomical Association Lunar Section (BAA http://website.lineone.net/~jmhh/index.html) and the American Lunar Society (ALS http://otterdad.dynip.com/als/). They run special observing programmes for cataloguing lunar impacts, domes or lunar transient phenomena. But they also provide a lot of helpful information for beginners.

The movements of the Moon

*W*hole books have been written about the Moon's motion around the Earth, such is the difficulty of computing its position on its orbit. The Moon speeds up and slows down erratically as it is subjected to the combined attraction of the Earth, the Sun and, to a lesser extent, the planets. It takes formulae with several hundred parameters to describe these variations correctly.

Earth–Moon: a double planet

The Earth and the Moon are an exceptional combination in the Solar System. The Earth's diameter is not even four times that of the Moon. Only Pluto and its satellite, Charon, have a greater ratio than this, but they are two tiny bodies 1200 and 800 km across. Among the major planets, no other satellite is as large or as massive as the Moon in proportion to its parent planet. If we could view the two bodies from far enough away, they would look like a double planet in space. From Venus, when at its closest to us, the Earth would appear as a binary planet with the apparent separation of its components attaining 1° at most. Much of the Moon's orbit around the Earth could be tracked by the naked eye.

The lunar orbit

The Moon and the Earth form a system held together by attraction and it is this system that revolves around the Sun. It is the barycentre (also termed the 'mass centre') of this system that describes an elliptical orbit around the Sun, improperly referred to as the 'Earth's' orbit. But the Earth and the Moon themselves have sinuous orbits. The Earth can be said to snake along this orbit, as shown in the diagram on p. 17.

The Moon's revolutions

It is difficult to determine the position of the Moon around the Earth. In addition, it

THE FAR SIDE OF THE MOON

The Moon takes the same time to spin around its axis as it does to orbit the Earth, with both motions being in the same direction. The rotation is said to be synchronous. That is why the Moon always turns the same hemisphere towards the Earth and why that hemisphere is called the 'near side'. The opposite hemisphere, the 'far side', is never visible from Earth.

A FEW FIGURES FOR COMPARISON

	Moon	Earth
Diameter at the equator (km)	3476	12 756
Circumference (km)	10 920	40 000
Surface area (km²)	37 960 000	512 839 600
Density	3.34	5.51
Gravity (m/s²)	1.62	9.81
Volume (billion km³)	22	1083
Mass ($\times 10^{21}$ tonnes)	0.073	59.34

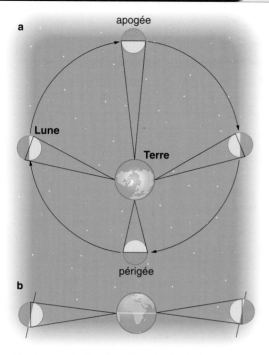

The two main librations (longitudinal above, latitudinal below) bring into view narrow stretches of the lunar far side.

FROM THE EARTH TO THE MOON

The average distance from the Earth to the Moon is 30.13 times the Earth's diameter. It is because they are so close that the interactions between the Earth and the Moon are so strong (tides).

As the Moon's orbit is elliptical, the distance Earth–Moon varies at each instant between the perigee (minimum distance) and the apogee (maximum distance).

To simplify, we can take the following values:
Minimum distance: 356 375 km
Mean distance: 384 408 km
Maximum distance: 406 720 km

The Earth and Moon are moving apart by 4 cm per year because of the dissipation of energy caused by the ocean's tides on Earth.

depends whether the Moon is to be positioned relative to the Sun or relative to the stars.

A 'synodic revolution' is relative to the Sun. It gives the time taken for the Moon to return to the same phase. It is the duration, for example, between two Full Moons. A 'sidereal revolution' is relative to the stars. It is the time taken for the Moon to return to the same celestial meridian after orbiting the Earth.

The Moon's changing faces

Everyone has noticed that the Moon seems to change shape as it moves around the Earth. This is not actually a change of shape but a variation in the way the Sun lights the Moon's orb. These are the phases of the Moon and are simply a matter of perspective that the diagram on p. 16 will help you to understand.

Completion of an entire series of these phases constitutes a lunar cycle, also termed a lunation or lunar month. It lasts the same time as a synodic revolution, i.e. 29.53 days.

Librations: lunar rock and roll

Librations allow observers to discover a narrow strip along the dark side of the Moon. Some 59% of the lunar surface is actually visible rather than the straight 50% that might be expected. The two main librations are:

• Latitudinal tilt (of 6°50′) caused by the inclination of the Moon's axis of rotation. This brings into view regions that lie just beyond the Moon's poles.
• Longitudinal wobble (of 7°54′) caused by the combination of orbital speed, which varies at different points along the trajectory, and rotational speed, which remains constant. The Moon travels faster on its orbit when it is at its closest to the Earth, its perigee, and more slowly when it is furthest away, at its apogee. But at the same time it turns on itself at a

THE MOON'S ORBITAL CHARACTERISTICS

Synodic revolution	29.53 days, i.e. 29 days 12 h 44 min 2.9 s
Sidereal revolution	27.32 days, i.e. 27 days 7 h 43 min 11.5 s
Direction of revolution	Normal (anticlockwise)
Average orbital speed	3683 km/h, or about 1 km/s
Apparent motion relative to the stars	33'/h or 13.176°/day on average

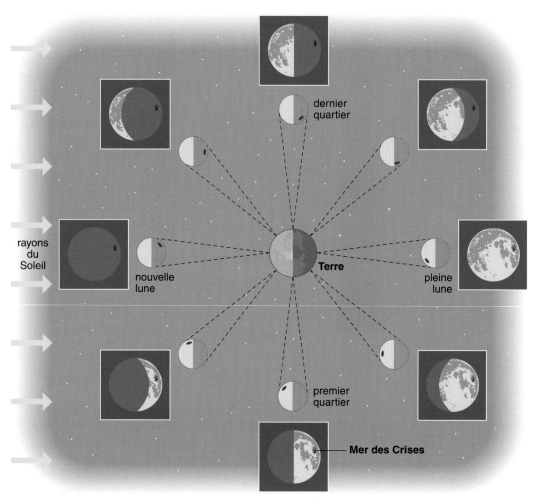

The phases of the Moon are simply variations in the way it is illuminated. Take the position of Mare Crisium (Sea of Crises) as a marker to see that the phases result from the Moon's rotation about its axis as it orbits the Earth and the direction of the Sun's rays. The boxes show the Moon as it appears to an observer in the northern hemisphere.

HOW TO RECOGNIZE THE MOON'S PHASES

Appearance	New Moon	First Quarter	Full Moon	Last Quarter
Position in sky	Near Sun	90° to Sun	Opposite Sun	90° to Sun
Rises	Dawn	Noon	Dusk	Midnight
Sets	Dusk	Midnight	Dawn	Noon
When visible	Invisible	Late afternoon and evening	All night and early morning	Second half of night

Here's an easy way to remember things if you know a couple of words of French:
At First (*premier*) Quarter, the Sun shines on the right-hand side of the Moon and we see what looks like the top part of the lower-case letter *p* as in *premier*.
At Last (*dernier*) Quarter, the Sun shines on the left-hand side of the Moon and we see what looks like the bottom part of the lower-case letter *d* as in *dernier*.

constant rate. So as it travels from perigee to apogee the Moon slows down in its orbit. At the half-way point, it will have spun around its own axis by less than a quarter of a turn but will have travelled exactly a quarter of its orbit. This allows an observer on Earth to glimpse a part of the far side of the Moon along its eastern edge. The process is repeated in reverse once the Moon has passed the

apogee; the Moon speeds up allowing part of the dark side of the Moon along its western edge to be seen from Earth.

These librations occur simultaneously but with different frequencies. Their combination in different ways enables 59% of the lunar surface to be visible at some time or other.

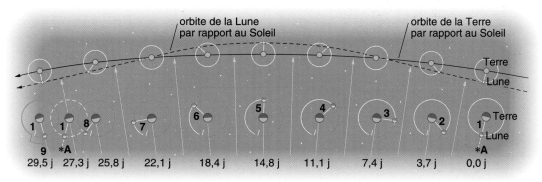

It is the centre of gravity of the Earth-Moon system that describes the orbit around the Sun. Note, on the left, the sidereal (between 8 and 9) and synodic (9) revolutions. (Distances to scale.)

Observation equipment and sites

The observer's equipment

Astronomical telescopes are not truly multi-purpose instruments and are often suited to a particular type of observation. The choice of equipment involves a number of criteria such as magnifying power or resolving power, the right type of mounting and your budget.

Choice of mounting

Altazimuth mounts, with two simple movements (horizontal and vertical), are not recommended for anything more than elementary lunar observations in the form of a quick look without detailed study. Computer-driven altazimuth mounts have two drawbacks: they are very expensive and they do not prevent the field of view from rotating slowly while tracking. To ensure the field of view remains fixed and to offset the motion of the Earth requires a clock-driven equatorial mount. A fork equatorial mount is the most convenient but a German-type equatorial mount is suitable too, especially if fixed on a post. Some equatorial mounts are fitted with a special Moon-tracking speed which slows the drive very slightly, correcting for the motion of the Moon relative to the stars. This is useful above all for high-resolution photography. A motor-driven axis of declination is a pointless luxury. This leaves the question of setting up the telescope. For a quick few minutes' observation it is good enough to line up the telescope by eye by carefully aiming the axis of right ascension at the Pole Star. For those wanting to explore the Moon for several tens of minutes or to take some good-quality photos an equatorial mount with a polar alignment scope is called for.

The ideal set-up is to have a permanent stand carefully aligned by the *two-star method* with the instrument remaining constantly under cover. And if the mount has lunar tracking

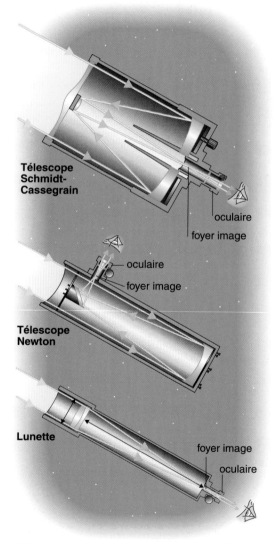

All three main types of telescopes are suitable for observing the Moon.

Telescopes with a star diagonal, like this refractor (left) or a Schmidt–Cassegrain telescope, invert left and right when viewing.

This 200-mm Schmidt–Cassegrain telescope needs a site with good seeing to give its best.

speed then it is worth thinking about some high-resolution photography . . . provided that the site is free from turbulence!

Choice of telescope

Any instruments, even binoculars and terrestrial telescopes or 'spyglasses', can be used for observing the Moon, but astronomical refractors and reflectors are the best.

Refractors are a good choice especially if they have apochromatic lenses, that is, lenses that correct the colour scattering caused when light passes through glass. This can only be done by using two or three lenses made of special 'ED' glass or of fluorite. But refractors like this cost much more than simple refractors with two 'achromatic' lenses. The

latter are adequate, though, since what chromatic aberration they do have does not interfere with the observation of details on the Moon.

Newtonian telescopes are suitable if their *focal ratio* is greater than *f*/8, if their tube is closed by an anti-reflective flat glass to reduce internal turbulence and if their mirror adjustment system is simple enough. Achieving these characteristics means modifying a telescope you have bought or making your own. The Dobsonian reflector is not very good for Moon viewing because of its open tube, its poorer quality lenses and its altazimuth mount.

Schmidt–Cassegrain or Maksutov catadioptric telescopes need to have an anti-reflective corrector plate, and the central obstruction caused by the secondary mirror should not be more than one-third of the corrector plate diameter. These telescopes are more sensitive to internal disturbance because the light ray is reflected twice inside the tube and so is subjected to any turbulence three times.

Criteria to consider

• **Optical quality.** Whatever type of telescope you use, the quality of the optics is paramount. For observing the Moon it is better to invest in a high-quality lens or mirror than in a tracking computer.
• **Alignment.** Even the finest optics will yield pitiful results if the components are not meticulously aligned. Refractors and Maksutov reflectors cannot be moved out of alignment as they are collimated by the manufacturer; even so the initial setting must be perfect. With a Newtonian telescope all the components can be adjusted, while on a Schmidt–Cassegrain telescope (SCT) only the secondary mirror can be adjusted. Both types must be adjusted regularly and if possible

before each observation session especially if the telescope has been jolted around in the back of the car.
• **Resolving power.** This is the capacity to discriminate between two adjacent details. It is calculated using the Dawes limit:

$$R = 120/\text{aperture (mm)}$$

where R is the resolving power in arc seconds. Unfortunately atmospheric *turbulence* makes images shake and very much limits a telescope's performance. The actual resolution is always dependent on the 'seeing', or turbulence in the telescope's line of sight (see p. 24).

In Britain, seeing on most nights is less than 1″, but theoretical resolution is sometimes achieved for a few tenths of a second. A few additional details may then be seen in a flash of clarity which is tiring to watch out for. Refractors add little internal turbulence to atmospheric turbulence because the light ray travels just once along the length of the tube. They are therefore at an advantage compared with reflectors when it comes to lunar

MAIN CHARACTERISTICS OF EACH TELESCOPE

Aperture (mm)	Low power	Medium power	High power	Resolving power	Diameter of smallest object theoretically visible on the Moon
50	30×	50×	100×	2.4″	4.8 km
60	40×	60×	120×	2.0″	4.0 km
70	50×	70×	140×	1.7″	3.4 km
80	60×	80×	160×	1.5″	3.0 km
90	70×	90×	180×	1.3″	2.6 km
100	80×	100×	200×	1.2″	2.4 km
120	80×	120×	240×	1.0″	2.0 km
150	80×	150×	300×	0.8″	1.6 km
180	80×	180×	300×	0.7″	1.4 km
200	80×	200×	400×	0.6″	1.2 km
250	80×	250×	400×	0.5″	1.0 km
300	80×	300×	400×	0.4″	0.8 km

observation, but apertures do not exceed 180 mm. The Dawes limit does not make allowance for contrast or for the shape of the features observed. That is why it is sometimes possible to make out a detail whose dimensions are below the resolving power, such as bright spots or very thin but elongate dark rilles.

• **Magnification.** This is determined by both the objective lens or mirror and the eyepiece. The classical formula for calculating magnification is:

$$M = F/f$$

where F is the focal length of the objective lens or mirror and f the focal length of the eyepiece.

If you want to observe the disc of the Moon in full so as to pick out regions for closer observation you can use magnification of up to about 80×. A practised eye can already spot many details.

Then to home in on the details that are theoretically visible with the telescope you need to switch to 'resolving magnification', which is equal to the aperture of the instrument in millimetres.

To see the details of lunar formations you will have to choose magnifications equal to the aperture of the telescope multiplied by a factor of 1.5–2.5. But beware! Such magnification can only be used when local seeing is better than the telescope's resolving power, which is just a few nights a year in Britain for apertures of more than 120 mm. In practice, then, three eyepieces are enough for observing the Moon: a low-power one for the whole disc, a medium-power one for fine details and a high-power one for maximum magnification of those details.

It is pointless having very wide-field eyepieces. Standard achromatic arrangements with four lenses, such as *orthoscopic* or *Plössl* eyepieces, with 31.75-mm push-fit barrels offer the best value for money for lunar observation.

To cut down on the number of eyepieces, you might like to use a Barlow lens on telescopes with short focal ratios. This doubles the magnification of each eyepiece although the price to pay is a few small flaws. So go for a three-lens apochromatic model with a 31.75-mm barrel.

• **Weight.** Lastly, your telescope should not weigh more than 40 kg if you are to carry it in one piece unaided.

And the budget?

If you have less than £1000 to spend, an 80–120-mm achromatic refractor or a Maksutov on a motor-driven mount is more useful for beginners than a 115-mm reflector.

For more advanced observation, if you have a budget of £1500–2500, a 200–250-mm Schmidt–Cassegrain telescope on an equatorial fork mount will allow detailed observation but still be transportable.

If you are lucky enough to have a permanent stand, for a budget of at least £5000, a 130–150-mm apochromatic refractor or a 250–300 Schmidt–Cassegrain, or even a Newtonian telescope with a 250–300-mm home-made flat glass will allow magnificent viewing or photography . . . if your site is right.

Where, when and how to observe the Moon?

*S*uccessful lunar observation involves a number of factors that determine image quality. The site is every bit as important as the time of observation or the method used.

Where? Almost anywhere

Need it be said that it is most inadvisable to observe through an open window, on a balcony or on a concreted terrace? Such places are exposed to swirling air and to the convection currents that produce *turbulence* and adversely affect seeing. Look for a more favourable spot around your home to reduce local turbulence. Ideally this will be on grass, in a garden, say, well away from any buildings because tarmac, cement and tiles store large quantities of heat during the daytime and give that heat out again in the evening. That is why expert lunar observers prefer to view in the morning, between Full Moon and New Moon,

but it means getting up early . . .

You will notice that the Moon can cope with the urban atmosphere. Unlike nebulae and galaxies, the Moon can be observed from the heart of a major city. Better still, atmospheric dust means that the images are often more stable than in the countryside. About 10 km from a large city you will find sites with moderate seeing. But, of course, there is nothing like the pure sky of the mountains. Are not the finest pictures of the Moon as viewed from Earth taken from high-altitude observatories?

When? Check local seeing

Before making observations, try to assess the local weather. Wind always causes strong turbulence that is highly detrimental to observations with a wide-aperture telescope, but it will not interfere with a glance through binoculars or a 60-mm refractor.

If there is no breeze to be felt, look at the brightest stars. If stars above 45° are shimmering, there is turbulence at altitude and you will usually have no chance of getting sharp images in anything more than

The Moon is the favourite target of urban observers, who are deprived of nebulae and galaxies by light pollution.

a 120-mm telescope. However, if only the stars at less than 45° from the horizon are twinkling, get your telescope out to judge local seeing. Seldom is there hardly any scintillation. Do not miss those nights . . . A slight mist often goes along with low turbulence but this does not detract from lunar observation because it acts rather like a filter.

There are three types of visible turbulence. The first is caused by the telescope, which you should bring out half-an-hour before beginning observation so that it can adjust to the outdoor temperature. This will eliminate any internal turbulence that causes severe 'boiling' of the images.

The second is caused by high-altitude disturbance. Images blur continually; only for a few tenths of a second does a sharp image form between two fuzzy episodes and it is difficult to search for fleeting details.

Turbulence spoils images, which remain blurred despite all attempts to focus.

The third type of turbulence is regular, as if the image were waving behind a curtain in a draught. Nights with such turbulence are common in Britain and good lunar

A SCALE FOR EVALUATING SEEING

Danjon and Couderc proposed a scale of seeing so as to indicate the conditions in which observations, drawings, images and photos are obtained. We have added the average frequency of turbulence in the countryside in Britain:

Level 1 – Perfectly motionless images
Theoretical resolving power achieved almost permanently.
Two or three nights *a year*.
Try to be there!
Highest magnifications on largest telescopes (2.5× or 3×).
Level 2 – Shimmering or slight waviness
Theoretical resolving power achieved for intervals of a few seconds.
Two or three nights a month.
Detailed observation at high magnification (2×).
Level 3 – Rapid shake with glimpses of finest

details
Theoretical resolving power achieved for a few tenths of a second separated by several seconds of blur.
About ten nights a month.
Detailed observation still feasible but very tiring.
Slightly higher than resolving magnification (1.5×).
Level 4 – Swirling and finer details masked
About ten nights a month.
Only general reconnaissance with no detailed observation.
Resolving magnification at best (1×).
Level 5 – Constant 'boiling' and so no observation
The other ten nights or so of the month, when you can catch up on lost sleep!

The pure air of the mountains and its low turbulence help lunar observation.

observations can be made. To minimise the effects of turbulence observe the Moon if possible when it is high in the sky. The First Quarter culminates in the evening in March, the Full Moon in the evening in December and the Last Quarter in the morning in September.

Lastly, we repeat that seasoned observers prefer the second half of the night as there is less turbulence, with observations being made between Full Moon and New Moon. But it is harder to get up early than to stay up late, especially when there is no guarantee as to the outcome! But none of these considerations will prevent you from applying the simple common sense solution: observe the Moon whenever it is out!

The best time: watch the terminator

The boundary between darkness and light on the Moon's disc is called the *terminator*. That is the line along which to observe and photograph. Along this curved line, the shadows of the least craters and hills stretch out exaggeratedly and the views are stunning. It is rather like the roughness of an old wall

that stands out clearly when the light falls on it at a low angle. The terminator changes position from one evening to the next and crosses the visible face of the Moon in 14½ nights (morning terminator, visible for us in the evening) and then, after Full Moon, in another 14½ nights (evening terminator, visible from Earth in the morning).

The best time to observe a lunar feature is when it emerges from the terminator (before Full Moon) or when it is about to pass into it (after Full Moon). The best periods for observations are during the three nights before and after the First and Last Quarters. The worst time is at Full Moon. The terminator shifts westwards at 15 km/h at the lunar equator. An hour after beginning observations, return to examine this area in detail because new features may now be illuminated.

How: be methodical

Fit the low-power eyepiece to your telescope. This is the time to note the direction and magnitude of the libration by examining the position of Mare Crisium (Sea of Crises) and of the craters Plato, Clavius and Grimaldi, depending on the phase of the Moon. If the libration is intense, the topography may look a little different from that described in this book. For example, part of a large crater may be visible although not mentioned here.

Begin by reconnoitring the places on the photos of the 'full' face of the Moon in this book for the night and the type of telescope you are using. Start at the top of the terminator and work down very slowly, identifying first the maria, then the mountains and lastly the major craters. Then locate the features that we have selected for you as being worth more detailed study.

Then change to the medium-power eyepiece. Find the first of the regions studied in the guide and try to discover the smallest details of each large feature as we describe them: the hills, craterlets, rilles and more rarely the domes.

Suppose you want to look at the Hyginus and Triesnecker rilles, which are not so easy to locate. Our chronological guide tells you that the best night to discover them is the seventh after New Moon. If you have a Newtonian telescope, the photo on p. 71 shows you that these rilles are in the centre of the lunar disc. If you have a refractor or a catadioptric telescope, look at the photo on p. 70. Use the low-power eyepiece showing the entire lunar disc and centre the region in question in the field of view.

Then switch to the medium-power eyepiece and locate the rilles using the photos of pp. 76–77. Adjust the focus of your telescope for a better view. The technique is to turn the knob until you are just past the right position, then to turn it slowly in the other direction so as to go past the focus as little as possible. Finally, turn it even more slowly in the first direction again stopping at the position that reveals the most detail.

If seeing is good, try the high-magnification eyepiece, but do not expect to find any new

A set of three Plössl or orthoscopic eyepieces is the best value for money for lunar observers.

detail. All this will do is magnify the detail you have already seen.

Make notes of the features you see, especially if you do not draw or photograph them. By referring back to your notes at each session you will be able to add to your knowledge of each feature that you study.

FROM MAP TO EYEPIECE

It is often difficult to find features plotted on a map through the telescope. The photos oriented for the type of telescope you are using as shown in this guide will help tremendously. But there may be some differences because of variations in lighting caused by librations.

Photographing the Moon: equipment

You will probably want to take photos of the Moon to show its wonders to your friends and share the fun of observation. Here are a few pointers to get you off to a good start.

Choice of telescope

First you will need a telescope with a motor-driven equatorial mount to offset the rotation of the Earth because exposure times are often between 0.5 and 4 seconds. 'Lunar' tracking speed is only useful if the telescope can be aligned precisely. Any type of telescope can be fitted to these mounts: a refractor, a Newtonian reflector or a catadioptric telescope. However, it must have an aperture of at least 80 mm so that there is enough light to keep exposure times short.

Remember that closed-tube instruments, such as refractors or catadioptic telescopes, generate less of the internal turbulence that is detrimental to image stability.

The ideal camera

You will need what is known as a single-lens reflex (SLR) camera, which uses a small movable mirror so that you see the same image through the viewfinder as appears in the field of view of the camera lens. Go for a 24 × 36 format that provides the widest range of film sensitivity. Your SLR camera must be operated by a cable shutter release as this prevents the camera and telescope from shaking when the shutter is tripped. The cable release must be at least 20 cm long, and nice and flexible. Your camera will also have to work on 'B' setting, that is to say that its shutter, when set to this position, remains open as long as you keep the cable release pressed.

A hand-held black card is needed to avoid mechanical shutter vibrations.

The focusing screen must be very finely ground so the camera can achieve a very sharp focus. If your camera has a focusing screen with Fresnel lenses, microprisms or a stigmometer, it will be difficult to use it for astrophotography. The ideal camera is an SLR with interchangeable focusing screens. Sadly those still on the market with these

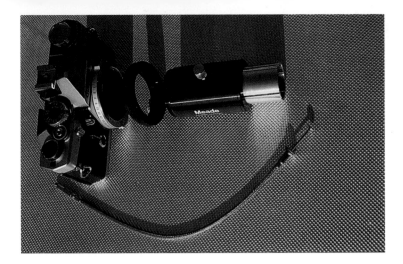

Everything for a good lunar photo: camera body, cable release, ring and photo adaptor tube.

characteristics cost the Earth! A third solution is to look for a second-hand one in the form of an Olympus OM1 or OM2, a Pentax LX, Nikon F, F2, F3 or F4, a Minolta XM, or a Canon F or F1. But be careful not to buy a camera unless you are sure you can find the right focusing screens and adaptor rings for it.

What type of focusing screen is best for lunar photography? The best is a clear glass with crosswires. 'Clear' means not ground and not with Fresnel lenses. 'With crosswires' means that the centre of the focusing screen features a simple cross.

Other requisite accessories

The camera now has to be connected to the telescope. Buy an astrophotographic adaptor tube designed for your telescope, preferably the one recommended by the manufacturer to be sure that it fits properly. Choose a model where the eyepiece can be fitted inside. To connect the camera to the photo adaptor tube you will need to find a T-ring with a female thread on the front to fit the adaptor tube and a male bayonet at the back to fit your camera. Finally, make yourself a hand-held blackout card, like a table-tennis bat half as wide again as your telescope's aperture.

Films

The aim is to achieve exposure times of 0.5 to 1 second, that can be achieved by hand, and that are short enough not to be troubled by atmospheric turbulence. This calls for sensitive films, but not overly sensitive ones or the grain will show up too prominently on any enlargements. A good compromise is to use ISO 400 Ilford and Kodak black-and-white films for paper prints and Agfa, Kodak and Fuji films for colour prints and colour slides.

Tips for successful lunar photography

Now you have all the equipment you need. You just have to apply two simple recipes. But your first sharp photograph will only come after a few failures.

Prime focus photography

This is the easiest technique and the best one to begin with. It can be used for photographing the whole lunar disc. The image of the Moon on the film will measure one centimetre for each metre of focal length; that is, a telescope with a 1.5 m focal length will produce an image 1.5 cm in diameter on the film. Beware: this technique cannot be used with some telescopes.

Step by step

1. First, get your telescope out half-an-hour before taking any photographs and use the time to line up the finder.
2. When the time is up, load a film into the camera and close the back. Wind the film on to view no. 1 and switch to 'B' setting.
3. Fit the astrophotographic focusing screen and the cable release on to the camera body.
4. Fix the T-ring on the bayonet. Screw the photo adaptor tube on to the T-ring.
5. That done, start the drive of the mounting and remove the eyepiece and star diagonal.

6. Fasten the adaptor to the telescope eyepiece holder. Check that its tube is balanced on the axis of declination and correct as required. If need be, mark the new setting on the tube so that it is easier to set up next time.
7. Get the Moon in the telescope's finder. It should be visible in the camera viewfinder too.
8. You can now start taking photographs. Turn

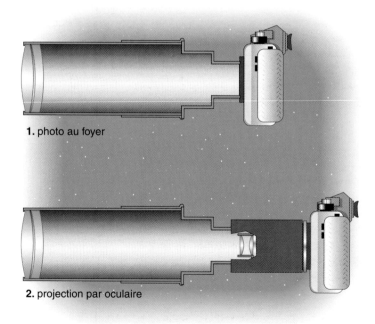

1. photo au foyer

2. projection par oculaire

The two principal techniques for photographing the Moon: prime focus and eyepiece projection.

To photograph the earthshine, use the prime focus technique with exposure times of 0.5 to 2 seconds with ISO 400 film.

the telescope focusing knob so that the Moon and the crosswires are both sharp.

9. Take hold of the cable release. Take the blackout card in the other hand and hold it in front of the aperture to cut out light to the tube. Trip the shutter and hold. After 5 seconds, when the vibrations disappear, move the hand-held card away and back so the tube is uncovered for about 0.5 second. Let go of the cable release button. There is your first photo of the Moon!

To be sure of getting a sharp photo repeat this operation ten times or so adjusting the focus each time. Try also to vary the exposure time from the shortest possible up to 1 second. There will always be one shot that is better than the others in the series.

Eyepiece projection photography

The procedure is just the same except that an eyepiece is inserted into the adaptor and

tightened before screwing the adaptor to the T-ring. Use an eyepiece with a focal length of 6–12 mm. Aim so that the region of the Moon you are interested in can be seen through the camera viewfinder and try a series of exposures of 0.5 s, 1 s and 2 s, adjusting the focus after each series. These attempts are almost certain to give a few acceptable photos out of a 36-frame ISO 400 film.

Your first photos will appear less detailed than the amateur photos in this book. But these lunar photography aficionados use apochromatic refractors or large-diameter telescopes and above all they have many years' experience. So keep at it and one day your photos will be up there with the best . . .

Electronic images: camcorders, webcams and CCDs

*F*or some years now, amateur astronomers have had access to new imaging techniques in the form of camcorders, CCD (charged-couple device) cameras and digital cameras. These techniques produce electronic images recorded in a digital medium. The advantage over classical photography is that the image can be easily touched up at home.

Using a camcorder

If you have a camcorder you can easily videotape the Moon even with an altazimuth mount.

Fit your camcorder on a photographic tripod. Set the zoom to two-thirds power. Set the camcorder objective lens manually to infinity. Fit your telescope with a 20–40-mm focal length eyepiece but without a star diagonal. Aim it at the Moon and bring it into focus. Centre the image on the region you want to film.

Then just place the camcorder lens centred behind the telescope eyepiece but not touching it. A distance of about 1 cm is a good compromise. If the camera is well aligned you should get a fairly sharp image in the viewfinder. Adjust the focus with the telescope's eyepiece holder and start recording. Adjust magnification by switching eyepieces rather than adjusting the focal length of the zoom.

Because of the Earth's rotation you can only film sequences of about 30 seconds. Film a few different regions each evening and you will be surprised at the results when you watch them on your television. If the clock-driven equatorial mount of your telescope can bear the weight of the camcorder, knock together a bracket to fix the camcorder to the telescope tube for reliable alignment and longer takes.

Until very recently affordable digital cameras could not be used for astrophotography because their lenses could not be removed. But the latest digital SLR cameras can be used just like conventional models and with equally effective results.

Newcomers: webcams

We are now into the realms of computing. Webcams are quite affordable, small CCD cameras designed for videoconferencing. They hook up to a laptop computer for greater convenience, although a desktop computer can also be used and is less expensive. These are no gadgets, though. A look at images obtained by users with 100-mm refractors will convince anyone of the value of these tiny cameras.

Remove the factory-fitted lens and infrared filter. Then bond or screw the webcam to the refractor with a T-ring. Connect the camera to the computer and launch the image acquisition software that comes with it. You can then obtain prime focus images or eyepiece projections by fastening the camera like an SLR camera on an photo adaptor tube. You control the focus directly on the computer screen and as there is no mechanical

A webcam fitted to at least a 90-mm telescope is an inexpensive way to begin in digital lunar photography.

Astronomical CCD cameras can achieve perfection . . . in skilled hands like those of Thierry Legault.

shutter, and so no vibration, there is no need for manual blanking-out and exposures are managed automatically by the software.

After the session you can use the image-processing software sometimes supplied with the camera to enhance contrasts and bring out finer detail.

The *ne plus ultra*: astronomical CCD cameras

Beware! A substantial investment is in store as these electronic cameras are expensive (£500–5000) and require special software as well as good understanding of how computer images are created and produced. In addition, these devices are designed for photographing nebulae and galaxies rather than the planets and the Moon.

However, a few brilliant amateurs have demonstrated that mastery of this tool can produce outstanding shots with resolutions of 0.40″, or 800 m on the lunar surface. These vie with the best photos ever taken from Earth.

Unlike webcams, most CCD cameras have mechanical shutters although they produce little vibration. The main difficulty is focusing by trial-and-error until the right distance is achieved. Using CCD cameras entails

investing in a laptop computer as well as an optical divider and a special reticule eyepiece for focusing and framing.

They are fairly complicated to use and we refer readers to specialist books for more detail. It is advisable to try your hand first with a webcam before investing in high-resolution lunar imaging with CCD cameras.

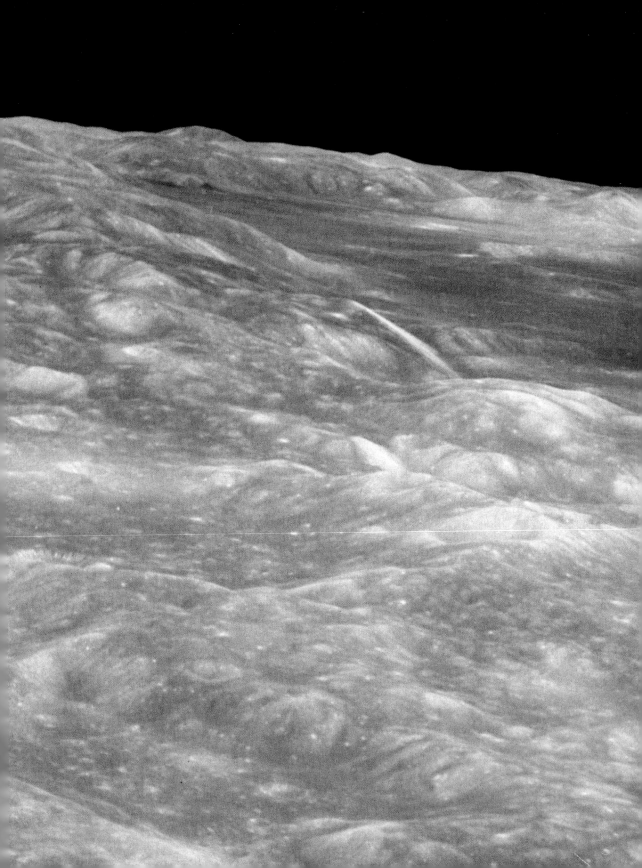

Guide to the Moon

Evenings 1 and 2

Yesterday was New Moon by your calendar. It is time for you to begin exploring our satellite. But your first challenge is to spot the young Moon, which has just started out on a new orbit of the Earth.

→ The Moon is still drowned in the Sun's radiation. Twenty-four hours after New Moon, our satellite makes an angle of only about 13° with the Sun. It is an extremely thin crescent, to be sought immediately to the left of the Sun as soon as it has set. A clear view of the western horizon is essential, and spring is the most favourable time of year. You will need binoculars to find the crescent which is scarcely any brighter than the background sky. The record for observing it is just 18 hours after New Moon! Beat it if you can.

→ The second evening after New Moon is more favourable for observation, with the crescent of Moon forming an angle of about 25° with the Sun. That is about the width of your hand with fingers spread, held at arm's length. You shouldn't have too much trouble

PATIENCE!

If you have two weeks of fine weather you will be able to track the sunrise over the face of the Moon each evening. Otherwise you will have to spread your viewing over several lunations, that is over several months. That is why we recommend that you observe the Moon whenever you can.

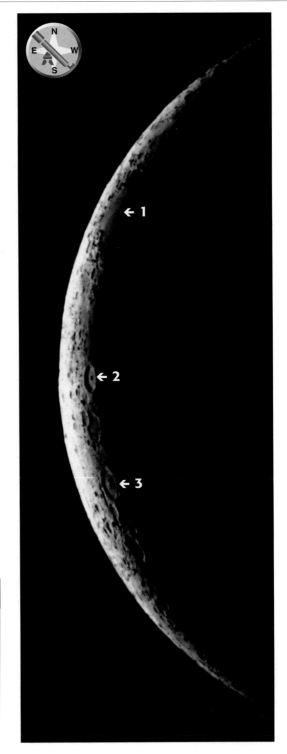

← 1

← 2

← 3

spotting this crescent. If you turn a telescope on it, you will begin to see a few features coming into view, as in the photos opposite.

Try to locate Mare Crisium (**1**), where the Sun is just rising, and a few craters viewed obliquely along the edge of the disc. The craters Langrenus (**2**) and Petavius (**3**) are likely to be illuminated already, but wait until tomorrow for a better view. Tonight is just for making acquaintance, for limbering up. Take the opportunity to check the set-up of the telescope, if it is an equatorial mount, and to practise focusing.

A SHORT LUNAR LEXICON

In addition to the official terminology of the International Astronomical Union, amateur astronomers use lots of colourful terms:
amphitheatre: synonymous with crater
crater-cone: dome with a crater atop it
craterpit: synonymous with craterlet
ghost: a formation drowned by lava
swelling: synonym of dome
monticule: small, circular mound
walled plain: large, ancient flat-floored crater
rays: streaks of material ejected from a crater
wrinkle ridge: synonym of dorsum

1. Mare Crisium
2. Langrenus
3. Petavius

Evening 3

*T*his evening you really begin to discover the lunar relief now that the Sun has been rising for three days on the eastern regions of the near side of the Moon. A mare and several magnificent craters are already on the agenda.

→ In the north, the only truly interesting feature is Endymion (**1**), a 125-km-wide crater whose flat floor is filled with dark lava and surrounded by ramparts 4600 m high. Then you will notice the group Messala (**2**), Geminus (**3**), Cleomedes (**4**) and the famous Mare Crisium (**5**).

→ Just beyond the equator, two near-perfect craters are prominent – Langrenus (**6**) and Vendelinus (**7**).

→ You can also observe Petavius (**8**), bordered by the Palitzsch Valley. South of Petavius lies a pair of craters that are much alike – Snellius (**9**), which is 83 km across with an uneven

OBSERVE ALONG THE TERMINATOR

The Sun has risen anew on the eastern regions of the lunar near side. You can make the most of the long shadows that enhance the relief of the Moon's features. This is particularly true along the curved line separating night and day (the inner boundary of the crescent) known as the 'terminator'. The terminator travels at 15 km/h and moves westwards each evening by 13°, or 350 km, allowing you to discover new regions night after night.

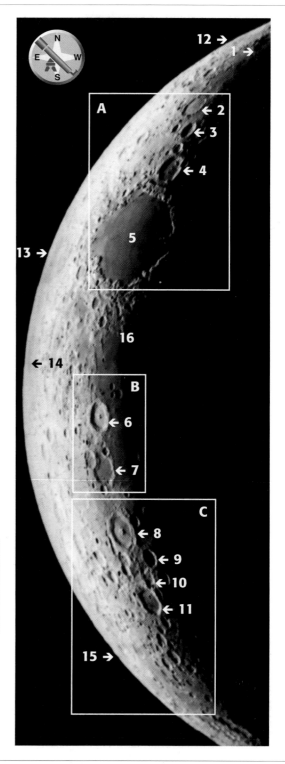

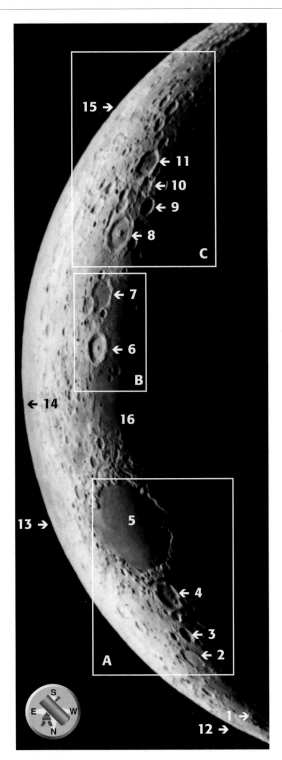

floor, and Stevinus (**10**), 75 km in diameter, ringed by a terraced sidewall and with a central peak. Then, just south of this twosome, you will find Furnerius (**11**), 120 km in diameter, whose floor is scarred by a rille and numerous craterlets.

→ Along the lunar limb you can also try to spot the four maria that lie mostly on the far side of the Moon, but which are sometimes brought into view by a favourable libration, namely, from north to south: Mare Humboldtianum (**12**), Mare Marginis (**13**), Mare Smythii (**14**) and Mare Australe (**15**).

Box A: *see* pp. 40–41

Box B: *see* pp. 42–43

Box C: *see* pp. 44–45

1. Endymion
2. Messala
3. Geminus
4. Cleomedes
5. Mare Crisium
6. Langrenus
7. Vendelinus
8. Petavius and Palitzsch Valley
9. Snellius
10. Stevinus
11. Furnerius
12. Mare Humboldtianum
13. Mare Marginis
14. Mare Smythii
15. Mare Australe
16. Mare Fecunditatis

→ Mare Crisium and its surroundings

A typical mare

Mare Crisium (**1**) is clearly visible throughout the lunation, even to the unaided eye, as a dark spot in the north-west of the lunar disc. It is an excellent indicator of the direction and intensity of lunar libration.

Mare Crisium clearly shows that lunar seas are in fact just gigantic craters whose floors were flooded by molten lava. Because of foreshortening, Mare Crisium looks circular when viewed from Earth, but it actually measures 570 km from north to south and 620 km from east to west.

It is ringed by the remnants of the ancient crater wall some 3000 m high. The lava filling the floor rippled as it cooled forming a wrinkle ridge which can be seen near the eastern edge. This lava also filled ancient craters within the basin, engendering the ghost craters Yerkes (**2**) and Lick (**3**), each some 30 km across. After the basin had filled with lava, a few more meteorites fell excavating Picard (**4**) and Peirce (**5**), some 20 km in diameter, and Greaves (**6**).

To the west, you can look for the Olivium (**7**) and Lavinium (**8**) promontories, separated by a valley some 10 km wide that is barred by a small hill. It was here, in 1953, that the amateur

astronomer O'Neil thought he had spotted a gigantic natural bridge 19 km long. In fact, it was just a trick of the light!

In August 1976, four years after the Apollo 17 mission, the probe Luna 24 bored to a

depth of 2 m beneath Mare Crisium, bringing the lunar exploration of the 1970s to an end.

Cleomedes and its neighbours

North of Mare Crisium lies Cleomedes (**9**), a crater that formed at about the same time as this mare. The crater wall was then smashed by the formation of several craterlets. Cleomedes is 126 km wide and 3000 m deep with a flat floor displaying a number of features – a small central peak, a very fine rille and a few craterlets. North of Cleomedes, the trio Burckhardt (**10**), Geminus (**11**) and Messala (**12**) is very interesting for the size gradation of these craters from 55 km, to 85 km and to 125 km, respectively. The terraced crater walls and the central peak of Geminus contrast with the flat floor of Messala.

1. Mare Crisium
2. Yerkes
3. Lick
4. Picard
5. Peirce
6. Greaves
7. Olivium Promontory
8. Lavinium Promontory
9. Cleomedes
10. Burckhardt
11. Geminus
12. Messala
13. Macrobius
14. Tisserand
15. Bernouilli

→ Langrenus and Vendelinus

Resplendent Langrenus

Langrenus is without contest one of the finest craters on the lunar near side. If it were located in the centre of the disc it would look like Copernicus. However, being situated close to the eastern limb, we see it much as would an astronaut in orbit around the Moon.

With a diameter of 130 km, Langrenus (**1**) has a very steep outer ejecta blanket giving on to a terraced inner wall some 2600 m high. Notice that the rampart is deformed to the south. The floor of Langrenus is not completely flat and it has hills about 100 m high. In the centre stands a mountain block with two peaks of about 1000 m. Langrenus is the centre of a ray system that can be seen clearly over the Mare Fecunditatis.

Observe too the interesting trio of craters Atwood (**2**), Bilharz (**3**) and Naonobu (**4**). Atwood is the smallest of the three at 30 km in diameter. Naonobu is 5 km wider and Bilharz 5 km wider again. These three craters have flat, lava-filled floors, and Naonobu alone has a craterlet in its surface.

An older crater – Vendelinus

Vendelinus (**5**) is 150 km across and is older than and very different from Langrenus. More recent

impacts have deformed the original crater. This is the case of Lohse (**6**) to the north, which is as deep as Langrenus but only 50 km in diameter, with a confined floor. To the east, Lamé (**7**), an irregular crater of 84 km diameter, has smashed the ancient rampart of Vendelinus.

The crater wall of Vendelinus is only low, at about 1000 m. It encircles an immense, flat floor pitted with craterlets, including a superb trio to the south made up of Vendelinus L, Z

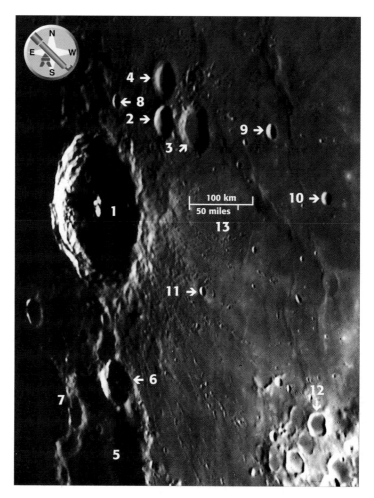

and Y. A few hills a hundred or so metres high can be seen to the south-west.

Notice too, to the north-east, a tongue of rocks that seems to have spread under the impact of Lamé, as if the original wall had been quite literally liquefied.

1. Langrenus
2. Atwood
3. Bilharz
4. Naonobu
5. Vendelinus
6. Lohse
7. Lamé
8. Acosta
9. Lindberg
10. Ibn Battuta
11. Al Marrakushi
12. Crozier
13. Mare Fecunditatis

100 km
50 miles

→ Petavius and the Palitzsch Valley

A real gem – Petavius

Petavius is the third of this evening's jewels. This crater, on the southern shores of Mare Fecunditatis, vies with Langrenus in interest, especially as it is flanked by the Palitzsch Valley.

The crater Petavius (**1**) is 177 km across and ringed by rugged outer slopes. These were smashed to the north-west by Wrottesley (**2**), a small crater 57 km across and 2300 m deep, with internal terracing, a confined, flat floor and a small central hill.

Notice that the outer slopes of Petavius rise above the 3300-m-high crater wall, which forms prominent terraces and, remarkably, splits into a double wall to the south. But it is the undulating floor of Petavius that is eye-catching. Many hills surround an imposing central mountain block some 30 km long with five peaks, the highest of which rises to 1700 m.

Now look in Petavius for a series of rilles, the largest of which appears as an 80-km-long line joining the central peak of Petavius to the south-west crater wall.

The Palitzsch Valley

Continuing your evening's exploration do not miss the Palitzsch Valley (**3**) on the eastern slopes of Petavius. It is formed by an alignment of at least seven overlapping craters.

The southernmost of these, at the end of the valley, is Palitzsch itself. The Palitzsch Valley is some 150 km long and as much as 40 km wide at Palitzsch, narrowing northwards.

Another crater of interest in this region is Furnerius (**4**). This is a walled plain some 125 km in diameter, the wall itself having been severely battered by several craterlets. The rugged floor features a number of

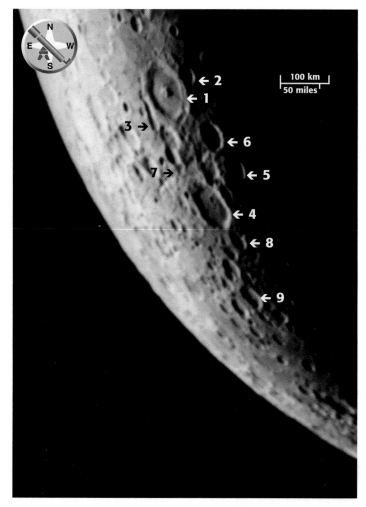

craterlets and, in its northern part, a rille. Finally, Stevinus (**5**) and Snellius (**6**), 80 km in diameter, are a contrasting pair worthy of attention. Stevinus has a flat floor with a fine central peak whereas Snellius has a more rugged floor. Notice also the Snellius Valley (**7**) which is at least 200 km long and 20 km wide, reminiscent of the Palitzsch Valley.

1. Petavius	**6.** Snellius
2. Wrottesley	**7.** Snellius Valley
3. Palitzsch Valley	**8.** Fraunhofer
4. Furnerius	**9.** Vega
5. Stevinus	

VALLEYS THAT AREN'T

On Earth valleys are produced by erosion mostly by rivers or glaciers and are downcut gradually. The Moon has no true valleys as there has never been any water to excavate them. What we call lunar valleys are actually either alignments of overlapping craters, or grabens, or broad clefts of volcanic origin.

Evening 4

*T*he crescent Moon, well away from the glow of the setting sun, clearly shows the earthshine. Try to spot the brightest features by it.

→ To the north of the crescent you will notice the pair of craters Hercules and Atlas (**1**). South of Atlas, Cepheus (**2**), a 40-km crater with a fine central peak, lies near Franklin (**3**), 56 km in diameter, with its 2700-m-high terraced walls and central mountain block.

→ On the western side of Mare Crisium (**4**) is Macrobius (**5**), 64 km in diameter, with its 3700-m-high terraced wall and its central peak. Beside it, Tisserand (**6**) has a flat floor which contrasts with that of Macrobius.

→ The eastern part of Mare Tranquillitatis (**7**) features the extraordinary Cauchy (**8**) region. Look too for Taruntius (**9**), 56 km in diameter, whose gently sloping 1200-m walls, smashed by Cameron, ring a rugged floor with a central peak and craterlets.

→ Mare Fecunditatis (**10**) measures 600 by 500 km, covering an area of 326 000 km².

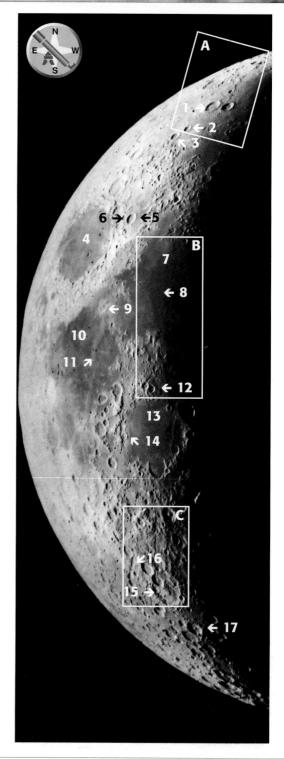

ADMIRE THE EARTHSHINE

While the crescent Moon shines, the remainder of the lunar disc can be made out in a faint, grey–blue light known as the earthshine. This is simply the sunlight reflected from Earth illuminating the Moon. The Earth shines 36 times more brightly than the Full Moon.

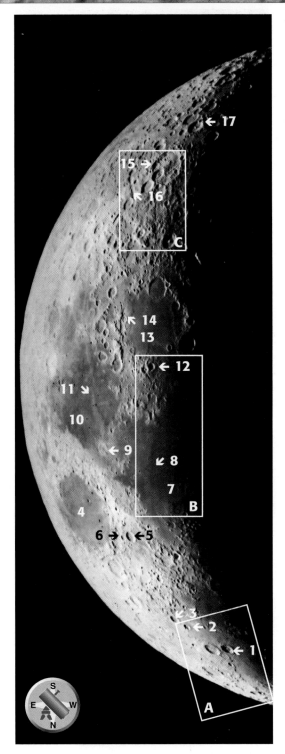

This irregular basin features the Goclenius Rilles on its western border and the exceptional twin craters Messier and Messier A (**11**).

→ A pair of craters some 40 km in diameter will draw your eye – Capella, which overlies Isidorus (**12**). A little further south, Mare Nectaris (**13**) is coming into view. This circular gulf of Mare Tranquillitatis is 350 km in diameter and covers an area of 101 000 km². Its rampart is formed by the Pyrenees Mountains (**14**).

→ The southern end of the terminator moves over an area with many old, overlapping craters that are difficult to identify. Do not miss Janssen (**15**) with the intriguing Rheita Valley (**16**) running to its east.

Box A: *see* pp. 48–49

Box B: *see* pp. 50–51

Box C: *see* pp. 52–53

1. Hercules/Atlas
2. Cepheus
3. Franklin
4. Mare Crisium
5. Macrobius
6. Tisserand
7. Mare Tranquillitatis
8. Cauchy
9. Taruntius/Cameron
10. Mare Fecunditatis
11. Messier/Messier A
12. Capella/Isidorus
13. Mare Nectaris
14. Pyrenees Mountains
15. Janssen
16. Rheita Valley
17. Vlacq/Rosenberger

→ On the eastern shore of Mare Frigoris

Atlas – a regular crater

Here is the first of the interesting crater pairs that dot the Moon's surface. Atlas (**1**) is the larger of the two with a diameter of 87 km. Its outer slopes are very even with many undulations and what looks like a large swelling of the ground to the east. The northern slopes extend as a straight mountain barrier 50 km in length. The 3000-m-high crater wall is steep despite the terraces that break the slope.

But it is the floor of Atlas that is its most singular feature. First it is littered with 100–200-m-high hills and looks as if it has been further excavated to the south. The central peak is merely a 300-m-high hill, only slightly higher than the others. A powerful telescope reveals a Y-shaped series of rilles some 80 km long but just 2-3 km wide, known as the Atlas Rilles.

Hercules

Hercules (**2**) is a complete contrast to its companion. Although at 69 km across it is smaller, it is none the less some 500 m deeper. Its flat floor has been flooded by dark lava. Hercules has a number of features including a craterlet 5 km in diameter and 800 m deep overriding the south-west crater wall.

Under favourable lighting you can see the landslip

detached from the northern wall and the 13-km crater that has pitted the floor of Hercules.

Other curiosities

North of the famous crater pair, on the edge of Mare Frigoris (**3**), Gärtner (**4**), a half-swallowed 100-km-wide crater ring, features a

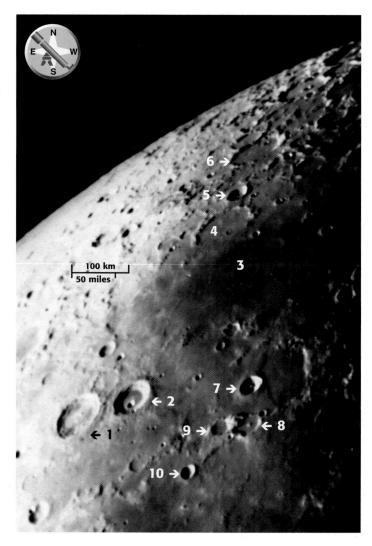

sinuous rille that is quite difficult to spot. Democritus (**5**), on the northern outer slope of Gärtner, is 40 km in diameter and 2000 m deep. It leads on to Arnold (**6**), a 100 km walled plain with a flat floor pierced by a 10 km craterlet.

The case of Mare Frigoris

While most lunar maria are roughly circular, testifying to their meteoric origin, Mare

Frigoris stretches more than 1000 km east–west across the northern part of the lunar near side. It is thought to be formed by a number of smaller, adjacent maria whose dividing walls were submerged by the final outflowing of lava that merged them into a single basin.

1. Atlas
2. Hercules
3. Mare Frigoris
4. Gärtner
5. Democritus
6. Arnold
7. Bürg and its rille
8. Plana
9. Mason
10. Grove

→ Cauchy and its region

An interesting cocktail

On the eastern edge of Mare Tranquillitatis, the same small area around the crater Cauchy includes three types of rare and therefore interesting lunar relief – a rille, a scarp and domes. Quite a mixed bag!

There is nothing exceptional about Cauchy (**1**) itself. It is a bowl-shaped craterlet some 12 km wide and 2600 m deep. But north of it is Rima Cauchy (**2**), a rille running north-west to south-east with an S-shaped section towards its western end. It is 120 km long and up to 4 km wide.

Looking south of Cauchy you will see Rupes Cauchy (**3**), a wall which is like a mirror image of the rille. This suggests that the two features are geologically related. Rupes Cauchy is a 120-km scarp, only 3 km wide, and flanked by a number of craterlets, one by the S-bend in the slope. This scarp is thought to be a thrust fault that formed as the lava that flooded Mare Tranquillitatis (**7**) cooled.

The Cauchy Domes

Now look south of the wall for two lumps in the ground – the Cauchy Domes. Cauchy Omega (**4**), which is 12 km at its base and rises to about 500 m, has an opening at its summit betraying its volcanic origin. Ash and gas probably issued from this vent a few billion years ago when the core of the Moon was still hot.

Further west, the second of the domes, Cauchy Tau (**5**), is as wide as its neighbour but higher. Its summit has no vent and it seems to

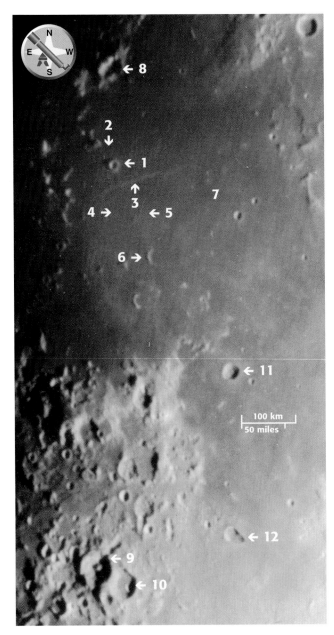

have been formed by a build-up of lava beneath the surface that made the surface swell without bursting. Use the horseshoe crater Aryabatha (**6**) to locate these two domes.

Mare Tranquillitatis (**7**) was a favourite target for the American lunar missions. Reconnaissance began in February 1965 with the crash-landing of Ranger 8. Then in September 1967 came the soft-landing of Surveyor 5 before the climax of the landing of Apollo 11 in July 1969.

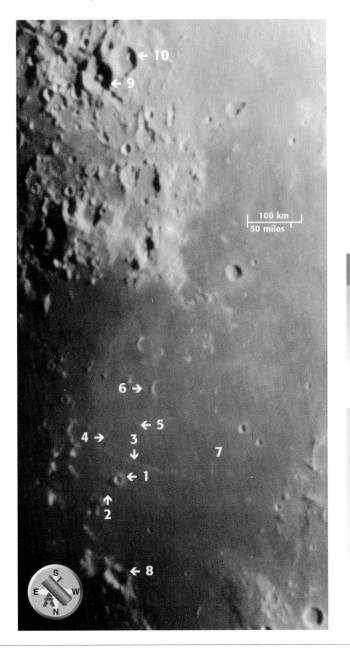

ELUSIVE DOMES

Lunar domes are difficult to observe. Their gentle slopes and altitudes of less than 500 m cast little shadow. They are therefore to be sought when the terminator is very close so that the shadows are at their longest.

1. Cauchy
2. Rima Cauchy
3. Rupes Cauchy
4. Cauchy Omega
5. Cauchy Tau
6. Aryabatha
7. Mare Tranquillitatis
8. Lyell
9. Capella
10. Isidorus
11. Maskelyne
12. Torricelli

→ The old crater and the valley

Janssen plain

Around 1960, Eugene Shoemaker developed a method of dating lunar formations on the principle that a new crater overlies older craters. You can test this theory by observing Janssen (**1**), a 190-km ruined walled plain. Notice first that its outer slopes vanish beneath numerous more recent craterlets. Its northern wall has been blown away by the double impact that created Metius (**2**) and Fabricius (**3**). The rugged floor of Janssen has many hills and a trio of craterlets to the east. A powerful telescope will pick out the network of rilles (**4**), the longest of which runs 140 km north–south.

Metius has steep outer slopes some 90 km wide. Its 3000-m-high terraced wall has been smashed by two craterlets to the north-west and rings a flat floor featuring Metius B and a central peak. Notice that the crater wall is dominated by that of Fabricius to the south, suggesting that Fabricius formed after Metius.

Fabricius (**3**) is smaller than its companion, at 80 km in diameter, and 500 m shallower. Its terraced wall is conspicuous and its flat floor has a central peak and an incredible straight range of mountains in its northern part.

Rheita and the Rheita Valley

North-east of Metius, the Rheita Valley (**5**) is a 500-km-long formation that reaches 30 km in width. It is formed by a line of about ten closely overlapping craters. This pseudo-valley begins south of Mallet (**6**), crosses Young (**7**) and seems to end between Metius (**2**) and Rheita (**8**), from which it takes its name. Closer scrutiny shows the valley extending south of Young. We now know that it was

AND WHAT ABOUT THE SOUTHERN HEMISPHERE?

When you are south of the Equator, the sky appears to be upside down compared with the view from the northern hemisphere. And so does the Moon! So people 'down under' may be wondering whether they too can use this guide? Well, yes, you can. Just turn your book upside down so the pictures match what you see through your telescope. Maps for the different types of telescope are unchanged. But you might get a stiff neck from trying to read the captions!

produced by the fall of enormous blocks catapulted in line at the time of the giant impact that formed Mare Nectaris further north.

Rheita (**8**) is an attractive crater 70 km across and 4000 m deep. Its terraced wall features a craterlet on the northern side and two on the southern side. A central peak rises over its flat floor. Notice also, to the north, Rheita E (**9**), an oddly elongate crater reminiscent of Schiller on a smaller scale.

100 km
50 miles

1. Janssen
2. Metius
3. Fabricius
4. Janssen Rilles
5. Rheita Valley
6. Mallet
7. Young
8. Rheita
9. Rheita E
10. Neander
11. Brenner

Evening 5

*T*his is an evening of transition in discovering the main lunar maria, with new ones coming into view while others are already entirely bathed in light.

→ The Sun is still rising over Mare Frigoris (**1**) and Mare Serenitatis (**2**). At the equator, Mare Tranquillitatis (**3**) is not yet completely illuminated. However, the 180-km-wide Sinus Asperitatis (**4**) connecting Mare Nectaris (**5**) to Mare Tranquillitatis (**3**) shows its floor covered in ridges and craterlets from the impact of the crater Theophilus.

→ Towards the north of the crescent, right on the eastern edge of Mare Serenitatis, you will notice the crater pair Posidonius/Chacornac (**6**) and the incomplete ring of Le Monnier (**7**).

→ Isolated at the end of Mount Argaeus (**8**), a 2500-m-high rocky headland near to which Apollo 17 landed, the 43-km-wide crater Plinius (**9**) has 2300-m terraced walls and fine rilles to the north. Its relatively flat floor has a central peak, landslips and two craterlets.

WATERLESS SEAS

The early observers of the Moon named the large dark patches that litter the near side 'seas'. These astronomers thought that the Moon behaved like a mirror reflecting the image of the Earth and so of its oceans. The maria are really vast, flat expanses of solidified basalt lava covering 31% of the visible hemisphere and were formed by liquid magma flooding over the surface.

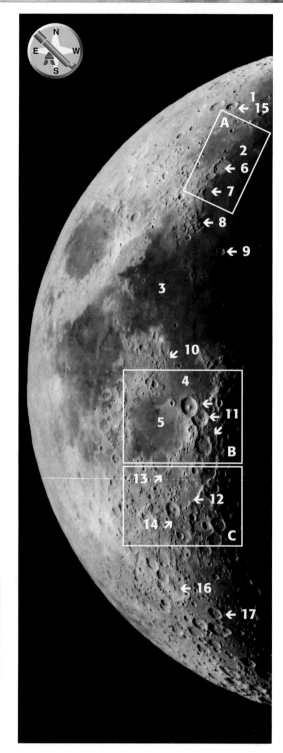

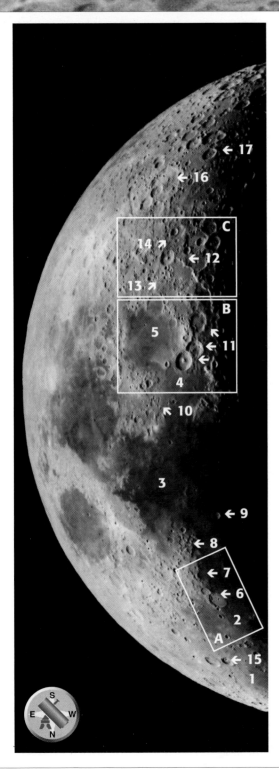

→ Within Sinus Asperitatis do not miss Torricelli (**10**), 23 km in diameter. This is a pear-shaped, double crater in the centre of a ghost crater. Its 2100-m-high walls ring a flat floor which connects up with that of its companion craterlet.

→ Just south of the equator is one of the marvels of the Moon – the three craters Theophilus, Cyrillus and Catharina – which you really must get a good close look at (**11**).

→ Continue your outing through Mare Nectaris (**5**), which is 350 km in diameter and covers an area of 101 000 km^2. It looks its 3.92 billion years' age as numerous impacts have degraded its circular shape. Its flat floor features two ghost craters near its northern edge. To the south, do not miss the Altai Scarp (**12**), Fracastorius (**13**) and Piccolomini (**14**).

Box A: *see* pp. 56–57

Box B: *see* pp. 58–59

Box C: *see* pp. 60–61

1. Mare Frigoris
2. Mare Serenitatis
3. Mare Tranquillitatis
4. Sinus Asperitatis
5. Mare Nectaris
6. Posidonius/Chacornac
7. Le Monnier
8. Mount Argaeus
9. Plinius
10. Torricelli
11. Theophilus, Cyrillus and Catharina
12. Altai Scarp
13. Fracastorius
14. Piccolomini
15. Hercules/Atlas
16. Janssen
17. Pitiscus

→ In the tracks of Lunakhod 2

A stunning feature – Posidonius

The most exceptional feature in this region is Posidonius (**1**), an ancient crater 95 km wide whose northern glacis has three craterlets. The crater wall is particularly rich – it is highest in the east reaching 1800 m and splits to form a mountainous crest that spirals inwards towards the crater centre. To the west, the wall dips progressively into the lava of Mare Serenitatis (**2**). The relatively flat floor of Posidonius features many craterlets, including the 10-km Posidonius A, which is almost central, and many hills rising between the rilles of Posidonius (**3**). These clefts are of major interest as two of them cross at right angles in the south-east and a third skirts the west wall.

Chacornac – a miniature Posidonius

Chacornac (**4**) is pretty much a scaled-down Posidonius. It is 51 km in diameter and contains similar landforms – a 1450-m-high, dilapidated crater wall, a 5-km central craterlet known as Chacornac A and Chacornac Rille which crosses the crater wall to the south-west. The crater floor is very hummocky.

Partly submerged Le Monnier

Le Monnier (**5**) was probably once a superb terraced crater before being drowned in part by the lava that formed Mare Serenitatis (**2**). The wall of this 61-km crater rises to 2400 m in the east but is visible as only a slight rise in the west.

Continue this outing by trying to find the remarkable, 15-km-long, G. Bond Rille (**6**) to the east of Posidonius and the Daniell Rille (**7**) to the west. Notice also the region

LUNAKHOD 2 EXPLORES LE MONNIER

Le Monnier became a historic place in January 1973 when the Russian probe Luna 21 set down Lunakhod 2 whose wheels were controlled remotely from Earth. The crawler was 2.2 m long and 1.6 m wide and carried three cameras, two of which were stereoscopic. Its wheels were made of wire mesh and its power supplied by solar panels on the 'cover' and by a small nuclear generator. Lunakhod 2 ranged over 37 km of rugged terrain in 10½ months, making many soil analyses and sending back 80 000 photographs and 86 panoramic images.

extending to the east and north of Posidonius known as Lacus Somniorum (**12**). End with the crater pair Plana (**10**) and Mason (**11**), which are of similar diameters though one has a central peak and the other a craterlet on a solidified lava floor.

1. Posidonius
2. Mare Serenitatis
3. Posidonius Rilles
4. Chacornac
5. Le Monnier
6. G. Bond Rille
7. Daniell Rille
8. Daniell
9. Grove
10. Plana
11. Mason
12. Lacus Somniorum

→ The magnificent three

Catharina, the oldest

Caution, dramatic landscape ahead! The trio Theophilus, Cyrillus and Catharina, three craters of the same size but different appearances, are among the marvels of the lunar surface.

Catharina (**1**), in the south, is the oldest of the three. This 100-km-wide crater has virtually no terracing left although its walls rise to more than 3000 m. It was smashed by the near-ghost crater Catharina P (**2**). Its floor is asymmetrical, being flat in the east and rugged in the west with many craterlets and hills.

In the middle, Cyrillus

Cyrillus (**3**) in the centre of the group is younger than Catharina as the terracing and the central mountain have been preserved. The 98-km crater was impacted on the western

1. Catharina	**10.** Bohnenberger
2. Catharina P	**11.** Rosse
3. Cyrillus	**12.** Mount Penck
4. Theophilus	**13.** Ibn Rushd
5. Sinus Asperitatis	**14.** Kant
6. Mädler	**15.** Isidorus
7. Mare Nectaris	**16.** Capella
8. Beaumont	
9. Daguerre	

side leaving a 17-km craterlet. The much-incised floor features a curved rille and a central mountain range with three peaks of decreasing size, the highest of which exceeds 1000 m.

Theophilus, the jewel

To the north, Theophilus (**4**), the jewel of them all, overlies Cyrillus proving that it is the youngest. The ejecta blanket in the north rose to 1200 m under the force of the impact. Theophilus is practically intact. None the less, some remarkable landslips can be seen to the south. Complex terraced walls some 5000 m high, that is, higher than Earth's Mont Blanc, lead down to the crater's flat floor with its many craterlets.

The mountain range in the centre of Theophilus is both spectacular and imposing. It extends over 30 km and peaks at 2000 m.

Three peaks, of regularly decreasing height, form a giant's stairway on the north side.

North of Theophilus, Sinus Asperitatis (**5**) is separated from Mare Nectaris (**7**) by Mädler (**6**). The Mare Nectaris basin contains two fine ghost craters, Beaumont (**8**) and Daguerre (**9**).

→ In the shadow of the Altai Scarp

Fracastorius

This splendid region lies to the south of the magnificent threesome described on pp. 58–59. Here again, three very different features are found in a small area.

On the edge of Mare Nectaris (**1**) lies Fracastorius (**2**), a 124-km crater whose north wall has been overwhelmed by lava which then spread over the crater floor. Inside, three tiny hills are all that remain of the central mountain, which implies that the lava is at least 1 km thick! Numerous craterlets dot the floor of Fracastorius.

Wonderful Piccolomini

Some 200 km south of Fracastorius an 85-km crater will catch your eye: this is Piccolomini (**3**). It is an imposing 4500 m deep, and a massive landslip to the south has brought billions of tonnes of rock on to the crater floor. The terraced crater wall towers over the floor with its many hills and a central mountain mass 2000 m high with several peaks.

CLIFFS THAT AREN'T

It was long believed that lunar 'walls' were sheer cliff faces. Examination of photographs from probes and measurement of shadows indicate that they are, in fact, transition zones between offset domains. Their slopes are far from vertical; indeed, they do not exceed 45°. You could walk down them with little trouble.

The fantastic Altai Scarp

End this evening's exploration with a long look at the tremendous Altai Scarp (**4**). This 480-km-long wall (about the length of England!) rises to an average altitude of 1000 m, reaching 3000 m in places. It starts at Piccolomini curving all the way round to Abulfeda. The high plateau is dotted with craters while at the foot of the scarp lies a strip of ground some 50 km wide devoid of craterlets. The Altai Scarp is probably all that remains of the original ramparts of Mare Nectaris.

At the foot of the Altai Scarp a powerful telescope will reveal the crater Polybius K (**5**). It is irregular in shape and its northern wall is straight for about 10 km, forming an extraordinary natural barrier a mere 250 m thick at its summit.

1. Mare Nectaris
2. Fracastorius
3. Piccolomini
4. Altai Scarp
5. Polybius K
6. Polybius
7. Weinek

A genuine impact basin

Mare Nectaris, 350 km in diameter and 101 000 km^2 in area, lies at the centre of an enormous impact basin 900 km across, part of the circumference of which is materialised by the Altai Scarp. The impact occurred 3.92 billion years ago.

Shortly afterwards the crater Fracastorius was formed. Then lava flooded the centre of the basin and many craters peppered this prodigious landscape.

Evening 6

One day before the First Quarter, and few interesting new features come into sight except for three remarkable highlights.

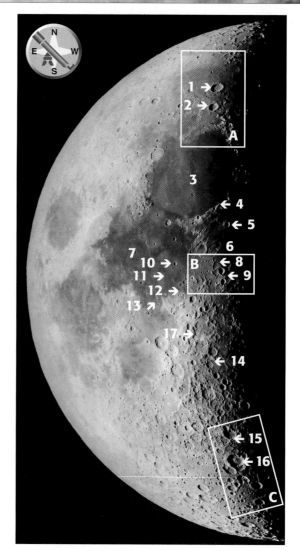

→ In the north, the crater pair Aristoteles (**1**)/Eudoxus (**2**) is prominent and is worth a good look around. The southern 'shore' of Mare Serenitatis (**3**), formed by the Haemus Mountains (**4**) and the crater Manilius (**5**), bears the scars of the impact that formed Mare Imbrium, which is still cloaked in darkness.

→ Amid the chaotic landscape between Mare Vaporum (**6**) and Mare Tranquillitatis (**7**), you can admire one of the finest lunar rilles – Rima Ariadaeus (**8**). South of this, have a look at the handsome pair of craters Agrippa/Godin (**9**).

→ On the western edge of Mare Tranquillitatis lies Arago (**10**), a marker for locating two rather inconspicuous domes. To the south-east, look for Lamont (**11**), a wonderful circular ghost feature visible only under very oblique lighting.

THE FIRST MEN ON THE MOON

On 20 July 1969, on the Apollo 11 mission, while Michael Collins remained orbiting in the command module *Columbia*, Neil Armstrong and Edwin Aldrin landed on the Moon in the LEM *Eagle*. On 21 July 1969 at 02.56 h (GMT), Armstrong first set foot on the lunar surface. For 2 h 31 min the two astronauts set up an ALSEP (Apollo lunar surface experimental package) including a seismometer and a laser reflector. They collected 21.4 kg of rocks over the course of the 21 h 36 min they spent on the Moon – hours that marked the history of humankind for all time.

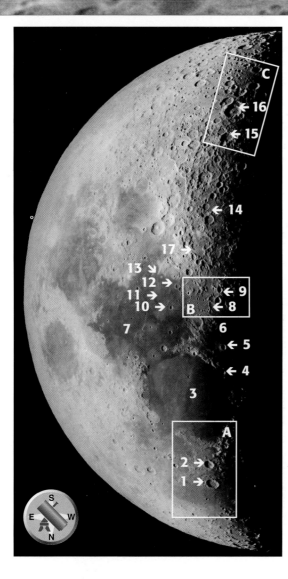

Lamont is a mystery because of the extensive network of wrinkle ridges that radiate from the ghost ring.

→ On the southern edge of Mare Tranquillitatis, 80 km from the pair Sabine/Ritter (**12**), two 30-km-wide craters, make a pilgrimage to the Apollo 11 landing site (**13**).

→ Further south, look for the Abulfeda craterlet chain (**14**). This line of craterlets 210 km long runs from Abulfeda ending north of the Altai Scarp. Continue your exploration by observing the complex of craters including Goodacre and Gemma Frisius (**15**) as well as Maurolycus and Barocius (**16**).

→ Lastly, you can also try to find the landing site of Apollo 16 (**17**) close to the crater Descartes, although the region is less interesting to observe.

Box A: *see* pp. 64–65

Box B: *see* pp. 66–67

Box C: *see* pp. 68–69

1. Aristoteles	**10.** Arago and its domes
2. Eudoxus	**11.** Lamont
3. Mare Serenitatis	**12.** Sabina/Ritter
4. Haemus Mountains	**13.** Apollo 11
5. Manilius	**14.** Abulfeda Chain
6. Mare Vaporum	**15.** Goodacre/Gemma Frisius
7. Mare Tranquillitatis	**16.** Maurolycus/Barocius
8. Ariadaeus Rille	**17.** Apollo 16
9. Agrippa/Godin	

→ The pair Aristoteles and Eudoxus

Aristoteles – a large crater

Its diameter of 87 km makes Aristoteles (**1**) as large as the famous Copernicus. Being located further north than Copernicus, it is viewed at a more oblique angle, making it a very attractive sight. Upon impact, rock was liquefied and sprayed out, forming rays. These streaks are still visible today, as is the 3700-m-high terraced wall, which has been subject to landslips to the south. The dark floor of the crater centre has small hills and two off-centre summits no more than 500 m high.

On the side of Aristoteles you will find Mitchell (**2**), a crater 30 km in diameter which is 1200 m deep with a rugged floor. From the lie of the land there is no telling whether Aristoteles overlies Mitchell or vice versa.

Eudoxus and its surroundings

About 80 km south of Aristoteles is Eudoxus (**3**), a crater 67 km in diameter which is more severely dilapidated than its larger neighbour. The steep outer slopes are uneven and have craterlets to the north. The terraced wall of Eudoxus is slightly lower than that of

1. Aristoteles
2. Mitchell
3. Eudoxus
4. Alexander
5. Lamèch
6. Egede
7. Mare Frigoris
8. Calippus
9. Galle
10. Sheepshanks
11. Mare Serenitatis
12. Caucasus Mountains

100 km
50 miles

ENORMOUS METEORITES

How big was the meteorite that formed a crater? A rule of thumb says that a meteorite forms a crater five to ten times its own size depending on its speed and its angle of impact. A crater 100 km in diameter is therefore the result of the impact of a block of rock between 10 and 20 km across.

Aristoteles and encircles a floor adorned with numerous hills.

South of Eudoxus a large field of hills extends as far as the old crater Alexander (**4**), which is badly dilapidated as it suffered a great deal from the impact that produced Eudoxus. The ancient, 100-km diameter crater lost its eastern ramparts. Its flat floor was filled with dark lava, which has cracked, and bears craterlets and hills, two of which are ringed by white material.

Egede and the Caucasus Mountains

West of Aristoteles is Egede (**6**). Its 35-km diameter walls now stand just 400 m above the lava of Mare Frigoris. Immediately south of Egede begin the Caucasus Mountains (**12**), a large range 500 km by 100 km whose 3500-m peaks are separated by deep valleys.

→ From Ariadaeus Rille to Godin Plateau

Ariadaeus Rille

If you own a small telescope do not miss the Ariadaeus Rille (**1**). It is easy to pick out with a 60-mm refractor and displays a mass of interesting details in telescopes with apertures of more than 100 mm. It was discovered in 1792 by Johann Hieronymus Schröter. This rille is without doubt a fault in the lunar crust, a graben between two cliffs. It is ancient because fresh relief has been created over it since it formed.

Measuring 220 km long and 4500 m wide on average it begins as a Y-shape in the east near the crater Ariadaeus (**2**), which is 11 km in diameter. The fault then runs westwards for some 40 km before being interrupted in several places by small hills. A further 40 km on, the fault seems to cut deeper and it vanishes west of the crater Silberschlag (**3**), 13 km in diameter, where a mountain range straddles it. The rille only truly reappears some 15 km further on before branching into a three-pronged fork to the west.

South of the rille

South of the Ariadaeus Rille, two young craters form a conspicuous pair. The more northerly, Agrippa (**4**), is 45 km wide and smashed the older Tempel (**5**) whose dilapidated wall rings a floor with many intricate features some 1200 m below. Agrippa is not quite circular. Its imposing 3100-m wall

has a number of terraces and a landslip to the south. A fine central peak rises from the flat floor.

South of Agrippa lies Godin (**6**), which is just as deep as Agrippa although smaller at 35 km in diameter. The floor looks to be hemmed in as the wall arrives at the foot of the central mountain. West of Godin, try to spot the strange 20 by 10 km Godin Plateau (**7**), which is a rare feature also visible in the photos on pp. 76 and 77.

This region has a wealth of bowl-shaped craters ranging from 10 to 20 km in diameter, such as Whewell (**8**), Cayley (**9**), De Morgan (**10**) and Dionysius (**11**).

Craters of this type are recent on the lunar geological timescale. They look very much like Meteor Crater in Arizona, the only well-preserved meteorite crater on Earth, except that they are five to ten times its size.

1. Ariadaeus Rille
2. Ariadaeus
3. Silberschlag
4. Agrippa
5. Tempel
6. Godin
7. Godin Plateau
8. Whewell
9. Cayley
10. De Morgan
11. Dionysius

→ In the maze of southern craters

Goodacre and Gemma Frisius

This evening we can zoom in on an intensely cratered southern region of the Moon. Several large, overlapping craters are eye-catching. First Goodacre (**1**) and Gemma Frisius (**2**). Goodacre is a strange hexagonal shape. Its 40-km-wide, 3000-m-high wall harbours a fresh crater. The flat floor has a low double mountain and a craterlet in the north.

Goodacre smashed the northern ejecta blanket of Gemma Frisius. This ancient, 90-km-wide crater has all but vanished in the bombardment it has undergone. But its great depth of 5000 m means it is still visible. Its flat floor bears landslips in the south and an off-centre hill.

Maurolycus and Barocius

Another interesting group is Maurolycus (**3**) and Barocius (**4**). Maurolycus is an old, 115-km crater with terraced walls still rising to 4000 m in the east. It formed over a yet older crater, part of which can be seen to the south.

EVEN FURTHER SOUTH

Do not hesitate to observe the Moon's South Pole. Standing out against the black shroud of darkness, rounded mountains await your visit. They are clearly visible in any telescope, making it hard to understand why jagged mountains survived in literature up until the Apollo missions.

A crater of 17 km diameter to the south and a jumble of overlapping craterlets to the north-west have overwhelmed the crater wall. The relatively flat floor features a few craterlets, one 8 km across, and a mountain range north of centre that links up with a landslip from the crater wall.

On the south-east outer slopes of Maurolycus, you will find Barocius (**4**). Its wall, ringing a floor 80 km wide , is adorned by two younger craters 30 km in diameter. The flat floor 3500 m below contains a ghost crater, hills, craterlets and a small off-centre mountain.

South of this pair, Clairaut (**6**) is worthy of interest. This old, 75-km crater had its wall blown away by two 25-km craters and its floor is studded by two 15-km craters.

1. Goodacre
2. Gemma Frisius
3. Maurolycus
4. Barocius
5. Buch
6. Clairaut
7. Breislak
8. Baco
9. Cuvier
10. Jacobi

Evening 7

*W*e are approaching First Quarter and you can look forward to a long evening's observation given the number of very varied features that are coming into sight.

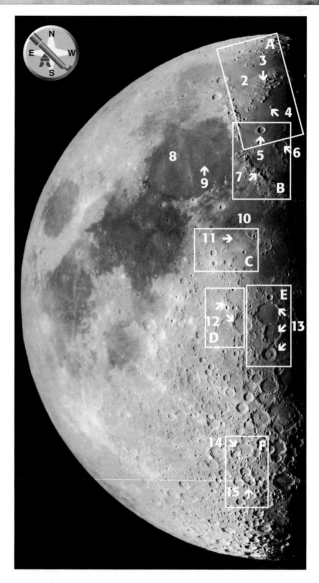

➜ The western shore of Mare Imbrium (**1**) is a rampart of mountain ranges. Below Mare Frigoris (**2**) you will notice the Alpine Valley (**3**), an enormous fault through the Alps.

➜ South of the Alps, your attention will be drawn by Mount Piton (**4**) and the Spitzbergen Mountains as well as by three very different craters – Cassini, Aristillus and Autolycus (**5**) – bordered by the Caucasus Mountains.

➜ Look next at Archimedes (**6**), Palus Putredinis and the majestic sight of the Apennine Mountains (**7**) with the Fresnel, Bradley and Hadley Rilles. The Apollo 15 astronauts landed close to Hadley Rille.

LINNÉ OR THE VAGARIES OF OBSERVATION

A number of nineteenth-century observers reported seeing changes in the size and shape of the crater Linné! They put it down as an active volcano and a long-running controversy about lunar activity waged until 1967 when photos from probes showed that Linné was only a recent 2.5-km diameter craterlet some 500 m deep. The explanation? As the Sun climbs higher over Linné it illuminates the 10-km ejecta blanket around the crater differently depending on the angle of illumination on each day of the lunation, the angle varying with the librations.

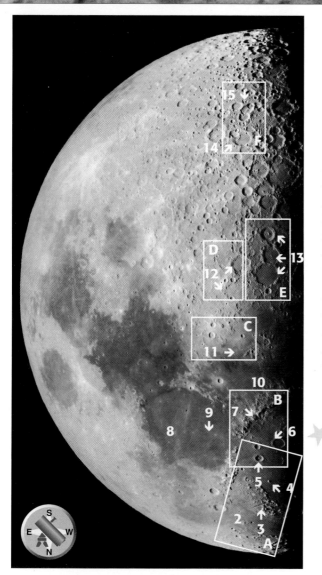

→ East of Hadley, in Mare Serenitatis (**8**), look for the small white patch Linné (**9**), a craterlet whose discovery caused quite a stir.

→ In Mare Vaporum (**10**), admire the splendid Hyginus Rille and the hairline network of the Triesnecker Rilles (**11**).

→ Further south, an interesting pair of walled plains, Hipparchus and Albategnius (**12**), await a visit as a literary salute to Tintin. The second, famous lunar trio of Ptolemaeus, Alphonsus and Arzachel (**13**) is strung out to the west of the previous pair.

→ South of Arzachel, look for Stöfler (**14**) and the crater complex surrounding it, including the strange grouping formed by Heraclitus (**15**) and Licetus.

Box A: *see* pp. 72–73

Box B: *see* pp. 74–75

Box C: *see* pp. 76–77

Box D: *see* pp. 78–79

Box E: *see* pp. 80–81

Box F: *see* pp. 82–83

1. Mare Imbrium
2. Mare Frigoris
3. Alpine Valley
4. Mount Piton/Spitzbergen Mountains
5. Cassini/Aristillus/Autolycus
6. Archimedes/Caucasus Mountains
7. Apennines/Palus Putredinis
8. Mare Serenitatis
9. Linné
10. Mare Vaporum
11. Hyginus Rille/Triesnecker Rilles
12. Hipparchus and Albategnius
13. Ptolemaeus/Alphonsus/Arzachel
14. Stöfler
15. Heraclitus

→ The valley of marvels

The Alps and the Alpine Valley

Just south of Mare Frigoris (**1**) lie the Alps Mountains (**2**), a range 250 km long and about 80 km wide with many peaks averaging 2400 m. The Alps are split into two very different parts by the Alpine Valley (**3**). To the west is a tangle of peaks separated by broad valleys. To the east, a vast range of hills is separated from Mare Imbrium (**4**) by a line of crests – Mont Blanc (**5**), the highest point of the Alps, a 25-km-wide mountain which rises to nearly 4000 m, and the Deville (**6**) and Agassiz (**7**) promontories. The Alpine Valley (**3**) is a sensational graben 130 km long and 11 km wide, bounded by 1000-m-high cliffs. Lunar probes discovered a 700-m straight rille stretching along its flat floor.

Cassini

At the end of the Alps Mountains you will come upon Cassini (**8**), a curious 60-km crater. Ringed by a wall no more than 1300 m high, with bulging outer slopes, its vast, flat lava-filled floor is pocked by two large craterlets 12 and 8 km in diameter.

Mount Piton

Do not miss Mount Piton (**9**), just west of Cassini. Its 25-km-wide base emerges from the lava to rise to 2200 m and its summit is pierced by an 800-m diameter crater. This isolated mountain is a vestige of an ancient inner ring of Mare Imbrium.

Aristillus and Autolycus

Aristillus and Autolycus await you next. Aristillus (**10**) is a perfect example of a lunar crater. It is 55 km across and its rugged outer

slopes rise to a terraced inner wall standing 3700 m above its flat floor, with a central mountain range with three 900 m peaks. Autolycus (**11**) is only 40 km in diameter but is just as deep as Aristillus. However, its floor features only a few hills. Look at the Spitzbergen Mountains (**12**) west of Aristillus. This straight mountain range is 60 km long and 15 km wide. It has four main peaks from 1100 to 1300 m high. To complete your excursion, examine the many bowl-shaped craterlets littering the region. Notice how the transition from a bowl-shaped craterlet to a normal crater occurs at about 20 km diameter, by observing the gradation of the 18-km Piazzi Smyth (**15**), 22-km Protagoras (**17**) and 25-km Theaetetus (**13**).

1. Mare Frigoris
2. Alps
3. Alpine Valley
4. Mare Imbrium
5. Mont Blanc
6. Deville Promontory
7. Agassiz Promontory
8. Cassini
9. Mount Piton
10. Aristillus
11. Autolycus
12. Spitzbergen Mountains
13. Theaetetus
14. Kirch
15. Piazzi Smyth
16. Trouvelot
17. Protagoras
18. Archytas
19. Timaeus
20. Egede

➜ On the trail of Apollo 15

Archimedes

The large crater Archimedes (**1**) forms a stunning trio with Aristillus and Autolycus. Archimedes is very different from its companions with narrower, steeper outer slopes rising to 2000 m. Inside the crater, terraces surround a superb, flat floor of dark lava 95 km wide with a few tiny craterlets on the western side.

The Apennines

South-east of Archimedes stretches the largest and most beautiful mountain range on the near side, the Apennines (**2**). The range is 950 km long and 100 km wide on average and has an impressive collection of lofty peaks. The highest of them is Mount Huygens (**3**) at 5500 m. Then come Mount Hadley (**4**) at 4800 m, Mount Bradley (**5**) at 4200 m and Mount Ampère (**6**) at 3000 m. Between Archimedes and the Apennines lies the Palus Putredinis (**7**). This is a flattish, square region with 180 km sides, containing the traces of ejecta from Archimedes.

Rilles in sight

This 'shore' of Mare Imbrium features several rilles that are rather difficult to pick out. To the north, try to find the three parallel Fresnel Rilles (**8**), 90 km long and 4 km wide. The hardest to observe is Bradley Rille (**9**), 130 km long and an average of 4 km wide. This rille was probably created by Mare Imbrium subsiding under the weight of lava that invaded it. But the most famous of them is Hadley Rille (**10**), visited by the

Apollo 15 (**11**) astronauts. This sinuous cleft is 80 km long and 2 km wide on average. It has a right-angle bend at its northern end and skirts the craterlet Hadley C at its centre. It looks very different to Bradley Rille. It is probably an ancient lava tube whose roof collapsed. The Apollo 15 mission showed that the inner parts of the rille slope at 45° and have curious strata of mysterious origin.

BY JEEP ACROSS THE MOON

Apollo 15 took David Scott and James Irwing, with Alfred Worden in the command module *Endeavour*, on the first 'high risk' mission to a mountain range. Scott and Irwing landed on 30 July 1971 in their LEM *Falcon*. With the first 'lunar jeep', during their 66 hours on the Moon, the two collected 77 kg of rock from depths of as much as 2 m, visited Hadley Rille, and left a seismometer and laser reflector.

1. Archimedes
2. Apennines
3. Mount Huygens
4. Mount Hadley
5. Mount Bradley
6. Mount Ampère
7. Palus Putredinis
8. Fresnel Rilles
9. Bradley Rille
10. Hadley Rille
11. Apollo 15
12. Conon
13. Galen
14. Aratus
15. Conon Rille
16. Archimedes Mountains
17. Spurr
18. Autolycus
19. Aristillus

→ Rilles galore

Hyginus Rille

A rille-feast this evening because, about 500 km south of the Appenines Rilles, the rilles of Hyginus and Triesnecker await your visit. The crater Hyginus (**1**), at 11 km in diameter with its 800-m walls dilapidated by a craterlet to the north, is not exactly spectacular. It is more interesting to observe the conspicuous rille that this crater overlies midway along its length.

Hyginus Rille (**2**) is 220 km long and an average 4 km wide. Broad where it begins in the west, it crosses Hyginus crater and then heads east and shallows. Its eastern portion joins up with Ariadaeus Rille (**3**) via a faint 40-km rille.

Through a powerful telescope, it looks as though it is formed from a line of craterlets. Its origin is mysterious. It is unlikely that clustered meteorites, even striking the ground in a straight line, could have given rise to such a feature. So is it a fault line along which collapse occurred? Only a visit to the site might elucidate its origins.

Triesnecker and its rilles

The other gem this evening is the network of rilles associated with the crater Triesnecker,

150 km south of Hyginus. Triesnecker (**4**) is barely more noticeable than Hyginus. It is 26 km across; its 2800-m walls are slightly terraced and surround a confined floor with a central peak. East of Triesnecker stretches the most stupendous network of rilles (**5**) on the lunar near side, covering an area 220 km by 50 km. This highly branched system has five main rilles which are joined by ten or so smaller clefts, sometimes at right angles. The most prominent rille, to the east of Triesnecker, is 3 km wide. These rilles originated in all likelihood as a network of underground tubes channelling the lava that formed the Sinus Medii (**6**).

1. Hyginus	8. Godin
2. Hyginus Rille	9. Godin Plateau
3. Ariadaeus Rille	10. Dembowski
	11. Tempel
4. Triesnecker	12. Boscovich
5. Triesnecker Rille	13. Ukert
	14. Chladni
6. Sinus Medii	15. Blagg
7. Agrippa	

➜ A pair of walled plains

Hipparchus, the crater Tintin made famous

South-east of Sinus Medii, a pair of dilapidated, old craters are worth a call. Hipparchus (**1**) is famous above all as the site where Tintin touched down in 1954 in *Explorers on the Moon*. Hipparchus is an ancient, eroded amphitheatre 150 km wide with outer slopes bearing numerous craters

including Pickering (**2**), Halley (**3**) and Hind (**4**). Its ramparts were disrupted to the south by Albategnius.

The crater wall is low in the north-west and on the east has two parallel valleys gouged by the impact of Mare Imbrium. On the floor of Hipparchus, Horrocks (**5**), 30 km across, is adorned with terraces and a central elevation. In the south is a near-ghost crater and many small 200–500-m hills dotting the plain. In

TINTIN ON THE MOON!

The lunar landscape imagined by Hergé in 1954 for his album *Explorers on the Moon* was consistent with the interpretations of astronomers of the time. Tintin moves among jagged mountains and steep-sided lunar craters. Admittedly some observers had evoked the possibility of gentle slopes and rounded mountains on the Moon, but many, judging from the appearance of shadows, thought the relief was sharp. Another difference is that Tintin explores a cave on the Moon. But caves form on Earth by water seeping through plateaux of sedimentary rock. So there can be nothing similar on the Moon. Lastly, Tintin finds ice at the bottom of a cave: a prediction that half came true . . . The probe *Clementine* discovered ice in 1995, but it was at the bottom of polar craters in perpetual darkness.

addition, near the western edge, hills form a circle, evidence of the presence of a buried crater.

Albategnius

South of Hipparchus, Albategnius (**6**) appears clearly younger. Its 120-km diameter terraced wall reaches a height of 4000 m. Smashed by a series of impacts to the north that produced a cluster of craterlets, it merges with Klein (**7**), a 44-km-wide crater, 1500 m deep with a terraced wall, flat floor and central peak. As with Hipparchus, the floor of Albategnius is filled with lava, suggesting that magma rose through the floor rather than flooding in from the Sinus Medii. The crater floor has several craterlets and a central mountain 2000 m high.

Finally, stop at an odd feature located on the south-eastern slopes of Albategnius. Vogel (**8**) is a 27-km-wide crater, 2800 m deep, at the centre of a chain of four craters. Together they look like a baby's rattle.

The entire region has a wealth of north-west–south-east trending valleys which are the remnants of material ejected when Mare Imbrium formed.

1. Hipparchus
2. Pickering
3. Halley
4. Hind
5. Horrocks
6. Albategnius
7. Klein
8. Vogel
9. Saunder
10. Müller
11. Ritchey
12. Burnham
13. Argelander
14. Airy
15. Réaumur Rille

→ An unforgettable trio

Ptolemaeus

Ptolemaeus (**1**) forms, along with Alphonsus and Arzachel, a trio that lunar observers sometimes confuse with the threesome Theophilus, Cyrillus and Catharina. It is an ancient ring 150 km across and reaching 3000 m in height in the south-west. Its outer slopes are adorned by a chain of craterlets to the north-east and two parallel valleys pointing towards Mare Imbrium. The dark, flat floor of Ptolemaeus was flooded by molten lava, leaving only ghost craters showing through. The most conspicuous is near Ammonius (**2**), but there are two more to the south. The floor has countless craterlets, some sixty of which are more than 1 km in diameter.

Alphonsus

The second of the trio, Alphonsus (**3**), is a hybrid crater 120 km across. Although as old as Ptolemaeus, Alphonsus looks younger because of its central peak which rises to 1000 m. The crater floor has many features including a pair of craterlets to the north, a rille on the eastern side and three dark patches on the edges that could be volcanic outfall.

1. Ptolemaeus	**7.** Spörer
2. Ammonius	**8.** Herschel
3. Alphonsus	**9.** Gylden
4. Arzachel	**10.** Müller
5. Aleptragius	**11.** Davy Y craterlet chain
6. Flammarion	**12.** Klein

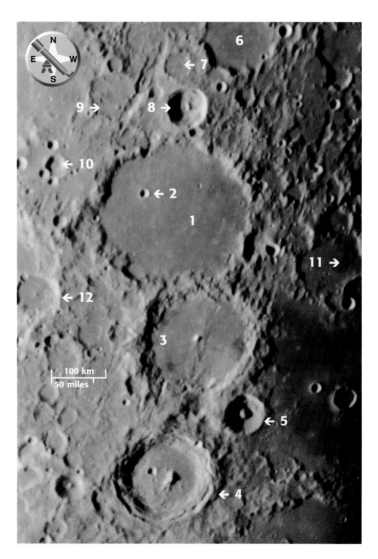

LUNAR TRANSIENT PHENOMENA – FACT OR FICTION?

Craters giving off clouds of gas . . . Many observers claim to have seen such events. There are few undisputed records. Only regular observation of Plato, Alphonsus and Aristarchus, which are constantly under scrutiny, might confirm the presence of residual volcanism. If you notice something untoward, try to photograph it, or better still film it, or at least draw it, showing its precise location. Then contact the Association of Lunar and Planetary Observers (ALPO).

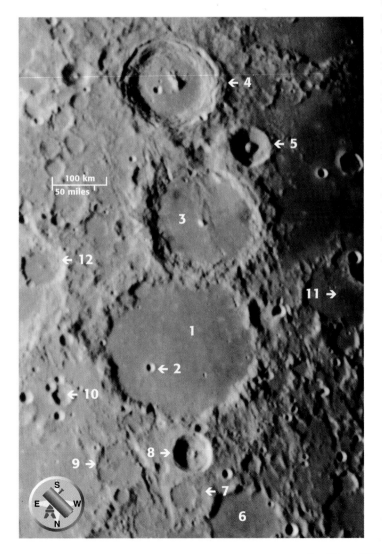

Transient phenomena have been reported in Alphonsus.

Arzachel and Alpetragius

Arzachel (**4**) looks the youngest of the trio. It is 100 km wide and its wall is terraced over its 3500 m height. The flat floor has hills alongside a 1500-m-high central mountain, adorned with a craterlet on its southern slope. The 11-km internal craterlet in the eastern part is helpful for finding a rille at the foot of the crater wall. Look a while at Alpetragius (**5**), an amazing floorless crater. Its terraced walls descend for 4000 m to abut directly on the foot of an impressive central cone 20 km across and 2000 m high. Finally, try to make out the chain of craterlets located in Davy Y (**11**).

→ A lunar battlefield

In this southern region of the Moon, the countless craters, petrified evidence of the intense initial bombardment to which it was subjected, all look alike.

Stöfler bombarded

Stöfler (**1**) is an impressive 125-km-wide crater whose 3500-m terraced wall is smashed in the south-east by Faraday (**2**). Molten lava flooded the floor of Stöfler which contains a few craterlets and elongated white patches that are probably ejecta from the crater Tycho.

   Stöfler was formed first. Two 40-km craters were superimposed on it in the south-east. Then a powerful impact produced Faraday (**2**), 70 km in diameter, the surface of which almost covered these two craters. Later Faraday was hit by Faraday P (**3**) in the south-west. Then, Faraday C (**4**) formed over Faraday P. And finally, Faraday A (**5**) was excavated to the north.

Most curious overlaps

Some 80 km south of Stöfler, another example illustrates the enigma of crater ages. Licetus (**6**), a 75-km crater, 3800 m deep with a central mountain, partly overrides the northern wall of Heraclitus (**7**). To the south, Heraclitus D (**8**) has covered the ancient rampart of Heraclitus. How can it be then that Heraclitus has preserved in the centre of its 90-km ring a very strange, straight mountain that extends to the wall of Licetus and which seems to have been formed after Licetus?

THE LUNAR GENERATION GAP

You will see craters of all ages on the Moon . . . Of the very old ones, all that remains is a faint circular framework visible at low Sun angles (e.g. Janssen). For more recent ones, the slopes have slipped or disappeared in places (e.g. Fracastorius, Gassendi). Young craters have sharp and fairly regular ramparts (e.g. Copernicus, Tycho). Finally, the freshest ones, which are usually the smallest, overlap the walls of their elders or pepper the maria.

Another interesting crater in the region is Lilius (**12**). Although of modest size at 60 km in diameter, it has a 1500-m-high central mountain poking through its completely flat floor.

1. Stöfler
2. Faraday
3. Faraday P
4. Faraday C
5. Faraday A
6. Licetus
7. Heraclitus
8. Heraclitus D
9. Kaiser
10. Fernelius
11. Cuvier
12. Lilius
13. Jacobi

Evening 8

This evening you are going to continue exploring, from the shores of Mare Imbrium to the great craters of the south. The day is just breaking over a few prime sites such as Copernicus and Clavius, but it is better to wait until tomorrow to examine them in detail.

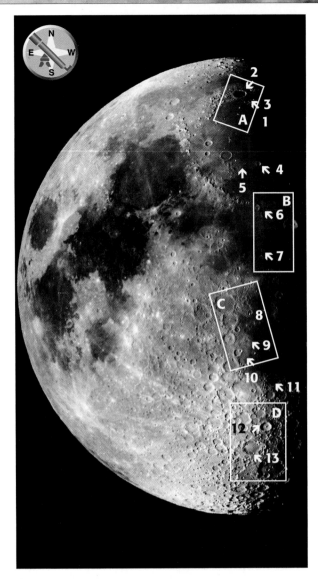

→ More than half of Mare Imbrium (**1**) is now illuminated and its circular shape is plain to see. On its northern edge lies Plato (**2**), also known as the 'greater black lake' because of its dark, lava-filled floor. To the south rises a stunning mountain range, the Teneriffe Mountains (**3**), with an isolated summit known as Mount Pico to the south.

→ To the west of Archimedes, look for Timocharis (**4**), a 34-km isolated crater with a ray system, and whose 3100-m terraced crater wall encloses a flat floor with central relief smashed by a craterlet. Between Archimedes and Timocharis, a handsome pair of 10-km craters, Feuillée to the west and Beer to

THE MOON'S ALBEDO

The albedo of a planet's surface is the ratio of the amount of light it reflects or diffuses to the amount striking it. The darker an object, the closer its albedo is to zero. The mean albedo of the lunar near side is 0.073. That of the maria is even weaker at 0.03, while the highlands are much brighter at 0.24. By way of comparison, the albedo of chalk is 0.85 and that of Mount Etna's lava 0.04. The Moon is therefore a comparatively dark body, reflecting only 7% of sunlight. If the Moon had the same albedo as Earth, the Full Moon would be five times brighter.

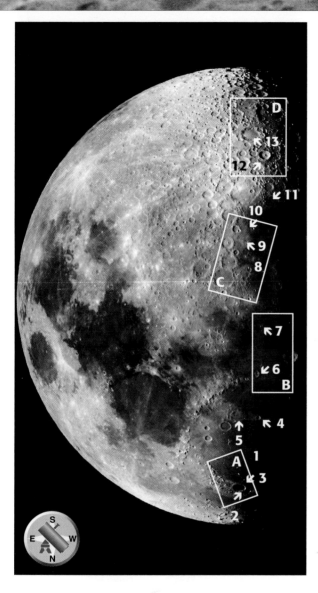

the east (**5**), are useful markers for an astonishing 40-km chain of craterlets on the southern rampart of Beer.

➜ At the western end of the Apennines is Eratosthenes (**6**), with a craterlet field to the west of it around the famous crater Stadius. South of Eratosthenes, the craterlet pair Gambart B and C indicates the position of two domes (**7**).

➜ Just south of the equator, Mare Nubium (**8**) is now clearly visible. This basin, measuring 750 by 500 km for an area of 254 000 km^2, is split into two similar parts by the Taenarium Promontory (**9**). It bears the dark scar of the Straight Wall (**10**), associated with Birt Rille, which is difficult to make out. On the southern edge of Mare Nubium, two craters attract attention: Pitatus and Hesiodus (**11**), which are neighbours of the old Deslandres.

➜ We end this evening's outing with the very young crater Tycho (**12**) and its impressive ray system. Tycho lies beside Maginus (**13**), an ancient crater with a rugged floor.

Box A: *see* pp. 86–87

Box B: *see* pp. 88–89

Box C: *see* pp. 90–91

Box D: *see* pp. 92–93

1. Mare Imbrium	**8.** Mare Nubium	
2. Plato	**9.** Taenarium Promontory	
3. Teneriffe Mountains	**10.** Straight Wall	
4. Timocharis	**11.** Pitatus/Hesiodus	
5. Beer and Feuillée	**12.** Tycho	
6. Eratosthenes	**13.** Maginus	
7. Gambart B and C		

→ The black lake

Plato, a majestic ring

This evening Plato (**1**) will draw your eye to the northern part of the terminator between Mare Frigoris (**2**) and Mare Imbrium (**3**). This large dark patch is an excellent indicator of the amplitude and direction of libration.

At 100 km in diameter, Plato is a glorious crater ring. Its dark floor contrasts with the light-coloured mountains that ring it. The 1000-m-high crater wall has a few 2000-m peaks that cast jagged shadows over the crater floor at sunrise.

Notice in the western part an impressive section of rock 20 km long which seems to have split from the crater wall. If you have a powerful telescope, try to spot some of the tiny craterlets that dot its floor. There are four that measure 2–3 km in diameter. To the north-west of the crater is a small, sinuous rille that is difficult to pick out (**9**).

THE MYSTERIES OF PLATO

Plato is remarkable on several counts. First, although it overlies the Alps Mountains, its floor is covered by the same type of lava as is found in Mare Imbrium. It is thought that Plato formed after the Imbrium impact but before the basin was flooded by dark, molten lava. Unless, that is, the impact of Plato burst an underground pocket of magma which then filled the crater . . . Intriguingly this solidified rock appears to darken as the Sun rises. Finally, Plato is the site of several transient phenomena (veils and clouds) suggesting that some tectonic activity persists at depth. It could be that pockets of gas seep through cracks in the ground.

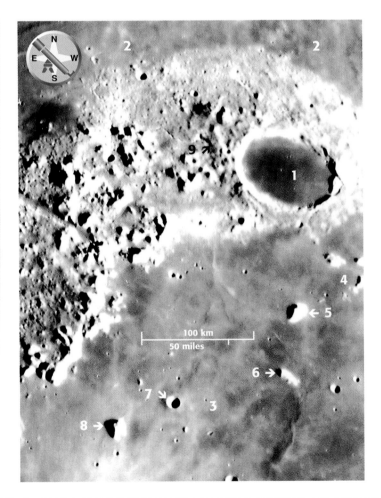

The Teneriffe Mountains and Mount Pico

South of Plato, look for the Teneriffe Mountains (**4**), a cluster of several bright isolated mountains emerging from the floor of Mare Imbrium. They are probably the remains of an ancient interior mountain ring that has been almost completely submerged. They rise to about 1500 m and stretch over a region 110 by 15 km.

Also south of Plato, look for Mount Pico (**5**), a lone mountain of 25 by 15 km. Its changing shadow is interesting to watch over the next two evenings, as it reveals the exact shape of the 2400-m-high peak and its slopes, which are no steeper than 5°.

Some 60 km south of Mount Pico stands another isolated mountain, referred to as Massif Beta (**6**) on some maps. Although as long as Mount Pico this massif is only 9 km wide and, at about 1500 m, is lower.

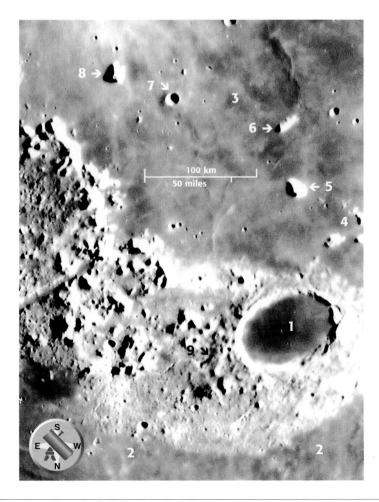

1. Plato
2. Mare Frigoris
3. Mare Imbrium
4. Teneriffe Mountains
5. Mount Pico
6. Massif Beta
7. Piazzi Smyth
8. Mount Piton
9. Plato Rille

→ A forest of craterlets

The stately Eratosthenes

The Apennines curve round to the south-west and the mountain range narrows considerably. At its end, the magnificent crater Eratosthenes (**1**) awaits your visit.

The 55-km crater rests against very steep outer slopes with most unusual undulations that extend southwards in the form of a short mountain range. The sharp, circular crest towers over terraced walls 4000 m high. The crater floor has many hills and, above all, an elongated central elevation with several peaks rising to altitudes of 2000 m.

Stadius – a strange feature

Just to the south-west of Eratosthenes lies a curious feature 70 km in diameter formed by a ring with craterlets and elongated hills. This is Stadius (**2**), the relic of a crater buried deep beneath the maria lava, the original rim of which barely emerges any longer. Numerous craterlets 1–2 km wide of mysterious origin dot this ring, some of them being elongate.

West of Eratosthenes, in the space between it and Copernicus, where the Sun is just rising, you can see an interesting chain of 1–3-km diameter craterlets (**3**) about 100 km long. It is hard to say whether these are impacts of rocks ejected by Copernicus or depressions excavated along a fault engendered by Copernicus.

A well-formed dome

Some 300 km south of Eratosthenes lies a pair of small, twin craters 12 km in diameter – Gambart B and Gambart C (**4**). A dome (**5**), with no crater at its summit, is found in the

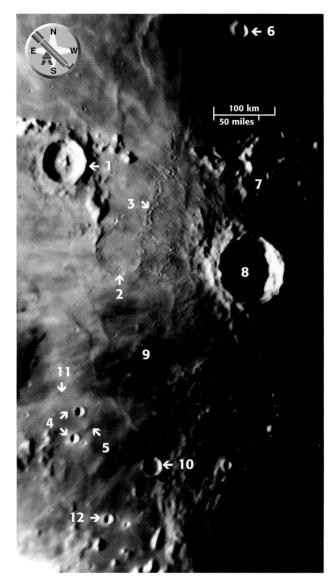

THE FABULOUS SUNRISE OVER COPERNICUS

As the terminator advances at 15 km/h, 6 h is enough to admire the sunrise over Copernicus. It makes for a marvellous night's observation. The Sun illuminating the loftiest parts of the crater is an impressive sight, as they appear to stand clear of the terminator, like stars. Wait a couple of hours and you will see the light joining the peaks to the lower-lying slopes. Wait a little longer and Copernicus will look like a simple ring of light. Finally, you will see the Sun flood the crater floor and discover the enormous central pyramid and its fantastic landslips. Six hours might sound a long time, but your patience will be rewarded by the beauty of the display.

vicinity. It is 13 km across but rises to barely more than 300 m. Because of its slight elevation, this volcanic feature is best observed at very low Sun angles.

1. Eratosthenes
2. Stadius
3. Stadius craterlets
4. Gambart B and C
5. Gambart Dome
6. Pytheas
7. Carpathian Mountains
8. Copernicus
9. Mare Insularum
10. Gambart
11. Surveyor 2 landing site
12. Turner

➜ The sword in the Moon

A sword-like fault

Between the craters Birt (**1**) and Thebit (**2**) you will explore one of the most amazing faults on the lunar surface – the Straight Wall (**3**), also known as the Sword in the Moon.

It is undoubtedly a fault line, more than 120 km long, running along a fracture probably engendered when the lava of Mare Nubium (**4**) cooled. Few commentators agree on its height and its gradient. Older books describe it as a cliff, while some of the more recent ones claim it is a gentle slope. High-resolution photography and determination of the day on which its shadow vanishes confirm that the Straight Wall is merely a 30–45° slope, some 1000–1500 m wide. Far from being the hoped-for paradise of future lunar mountaineers, the Straight Wall could be scaled easily by ordinary day-trippers.

East of the Straight Wall, a strange trio includes Thebit, 55 km in diameter and sinking to more than 3000 m, Thebit A (18 km in diameter), which overlies Thebit's terraced wall on the western side, and Thebit L, which in turn has smashed the rim of Thebit A.

Birt Rille

West of Birt, try to pick out a delicate north–south rille (**5**). This 70-km-long valley is no more than 1500 m wide. It seems to run down from a low dome at its northern tip,

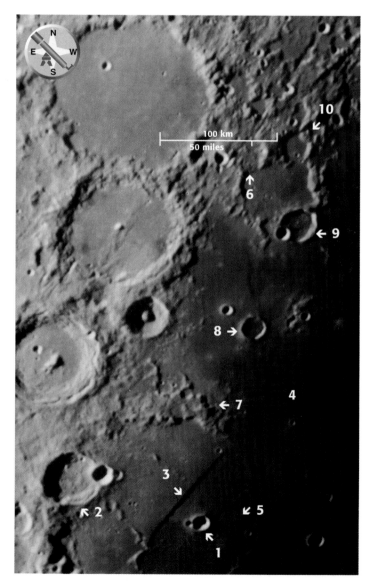

where it looks like an elongated crater. It is probably an ancient lava channel that carried lava flowing from the dome.

This is probably some of the best-preserved evidence of past volcanic activity on the Moon, along with Schröter's Valley.

In the same region, stop near Davy (**6**). Admittedly, it is not a particularly impressive crater, but its 35-km enclosure overlies an older crater that holds a treasure-trove – a string of a dozen 1–3-km craterlets stretching for 50 km. This remarkable formation was probably produced by the impacts of debris from a single meteorite.

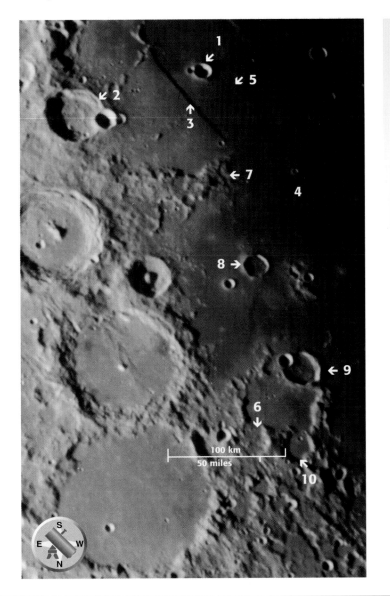

1. Birt
2. Thebit
3. Straight Wall
4. Mare Nubium
5. Birt Rille
6. Davy Y and its crater chain
7. Taenarium Promontory
8. Lossel
9. Davy
10. Palisa

→ The dinosaur crater

The brilliant, young Tycho

About 109 million years ago, when dinosaurs roamed the Earth, a 10-km meteorite struck the south of the Moon at high speed. Those great reptiles probably witnessed an exceptional conflagration of the lunar disc that gave birth to the crater Tycho (**1**).

The amazing brightness of this 85-km diameter crater proves how young it is. It seems to have been punched clean out of the highland region around it.

Tycho's intact wall has several rows of terraces soaring to 4800 m. Its floor is scattered with small hills surrounding a 10-km diameter central mountain which stands 1500 m high and has three main peaks.

Tycho is the centre of the largest ray system on the lunar near side. These spokes are particularly conspicuous at Full

Moon and are made of material thrown up by the impact of Tycho falling back to the Moon's surface. One of these streaks runs for more than 1000 km across Mare Tranquillitatis. The impact was certainly of tremendous proportions although the weak lunar gravity also explains this phenomenon.

Tycho is surrounded by a ring of older craters. Street (**2**) measures 58 km in diameter and looks very dilapidated compared with its

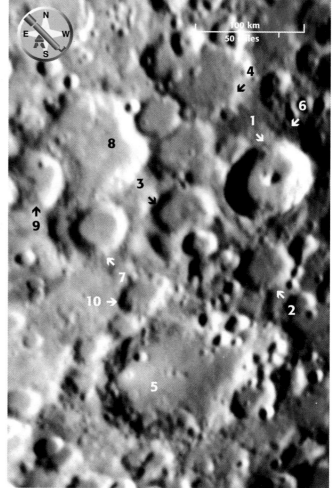

THE END OF THE DINO-SAURS

Sixty-five million years ago the dinosaurs suddenly vanished. There are two opposing theories at present: the theory of intense terrestrial volcanic activity or the theory of a giant meteorite strike that caused turmoil in the Earth's atmosphere. While the first hypothesis cannot be dismissed, the presence of Tycho on the Moon is a reminder that large asteroids were still cruising in the vicinity of Earth at that time.

TYCHO AND CLAVIUS AT THE PICTURES

If you want to feel you are walking inside Clavius and Tycho, get a copy of Stanley Kubrick's film *2001: A Space Odyssey*. Arthur C. Clarke, who wrote the screenplay, set his permanent Moon base in Clavius. But it was in Tycho that the scientists discovered AMT1, the black monolith, a sentinel buried several million years earlier by aliens. The film sets give a good impression of what craters are like. It is pointless looking for AMT1 with your telescope, though . . .

young neighbour. Its worn outer slopes do not exceed 1500 m and its rugged floor has a craterlet in its north-western part. Pictet (**3**), at 62 km, is larger than Street but has a wonderful group of five hills in its centre, which is 2700 m below its rim. Sasserides (**4**) is probably the oldest of the three, as its outer slopes are pocked with craters, making it difficult to pick out despite being 90 km wide.

Maginus, an old crater

South of Tycho you can explore the old, 170-km diameter crater Maginus (**5**). It is an interesting test for your telescope as its floor, ringed by a wall 4000 m high, contains many craterlets 1–5 km in diameter. In addition, six 1000-m-high central mountains extend for over 30 km.

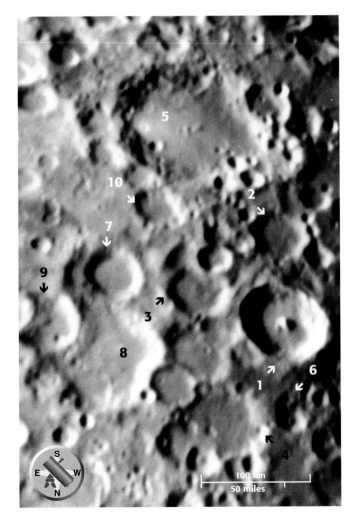

1. Tycho
2. Street
3. Pictet
4. Sasserides
5. Maginus
6. Surveyor 7 landing site
7. Saussure
8. Orontius
9. Huggins
10. Proctor

Evening 9

*T*he Moon is now properly *'gibbous' and you will continue to discover stretches of Mare Imbrium and Mare Nubium. But a few exceptional features will catch your eye above all.*

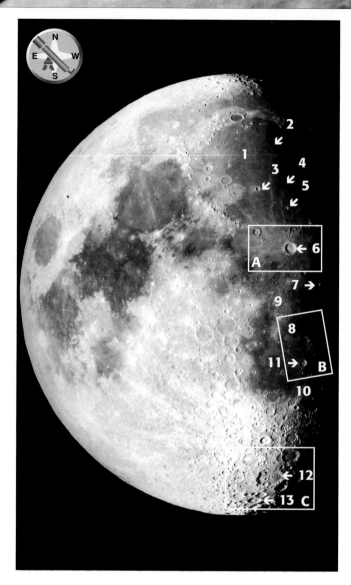

→ In Mare Imbrium (**1**), Helicon is 25 km wide with walls soaring 1900 m above its flat floor and central hill. It forms an interesting pair with Le Verrier (**2**), a slightly deeper crater with its 2100-m relief and larger central peak.

→ South of this pair, three separate craters await. West of Timocharis (**3**), a crater surrounded by bright streaks, you can see Lambert (**4**), which is almost a replica of Timocharis. Try to spot Lambert R, a ghost crater 40 km wide just south of Lambert, and observe Pytheas (**5**), a bright 20-km diameter crater 2500 m deep.

→ But the magic moment has arrived because you are at last able to admire the superlative Copernicus (**6**), with the Carpathian Mountains to the north and Mare Insularum to the east. South of Copernicus, do not miss

Reinhold (**7**), 48 km in diameter with its 3200-m-high terraced walls and a flat floor containing hills and craterlets.

→ This evening a new mare basin comes into sight: Mare Cognitum (**8**), bounded to the north-west by the Riphaean Mountains near to which Apollo 12 landed on 19 November 1969. And it was on the northern slopes of Fra

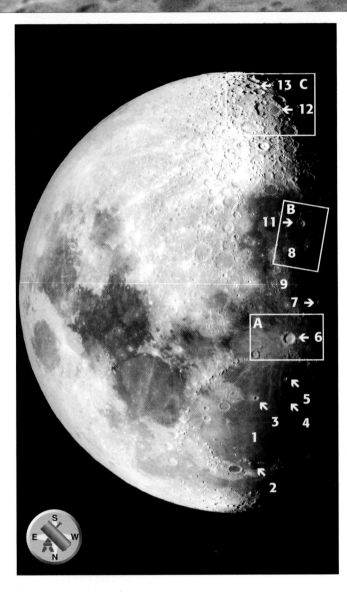

Mauro (**9**) that Apollo 14 touched down on 5 February 1971.

➜ South of Mare Cognitum, in Mare Nubium (**10**), you will find the splendid crater Bullialdus (**11**). Look closely too, to the south of it, for the ghost crater Kies and its famous dome.

➜ Finally, you will be able to see the enormous crater Clavius (**12**) and its companion Longomontanus, where a wealth of small features will appear to the persevering observer. Near this imposing pair, the magnificent crater Moretus (**13**) will provide an unforgettable view to end your evening's travels.

Box A: *see* pp. 96–97

Box B: *see* pp. 98–99

Box C: *see* pp. 100–101

1. Mare Imbrium	**8.** Mare Cognitum/ Riphaean Mountains
2. Helicon/Le Verrier	
3. Timocharis	**9.** Fra Mauro/Apollo 14
4. Lambert	**10.** Mare Nubium
5. Pytheas	**11.** Bullialdus
6. Copernicus/ Carpathian Mountains	**12.** Clavius/ Longomontanus
7. Reinhold	**13.** Moretus

→ The magnificent Copernicus

The finest crater

Copernicus (**1**) is also known as the 'Monarch of the Moon' and is without a doubt the finest crater on the near side. Its central location allows you to view all the components of large lunar crater in detail 'head on'.

Samples brought back by the Apollo missions show that Copernicus was formed 810 million years ago, making it 700 million years older than Tycho. That it is so remarkably well preserved proves that there has been little meteor activity since its formation. Even so, the presence of a few craters, notably Fauth (**2**), a double crater 20 km long and 2000 m deep, shows that there has been some bombardment.

Because Copernicus formed in a basin, it differs in aspect from Tycho. The ejecta blanket, 900 m higher than the surrounding maria, contains many radial valleys, and chains of craterlets can be seen all around the crater. The ejecta has been spread like a cobweb over an area of 500 km in diameter, whereas ejecta from Tycho shot off like artillery shells in straight lines over distances of more than 3000 km. It can be deduced from this either that the impact of Copernicus was less violent than that of Tycho or that the ejected maria material was more fluid.

The crater of Copernicus is like a hexagon 93 km wide. The wall collapsed to form a series of terraces over 4000 m high. Oddly, the crater floor is flat in the north whereas many small hills are scattered over the south. In the centre, the mountain massif stretches for 30 km and comprises three main peaks with gentle slopes rising to 1200 m, well below the crater rim.

The Carpathian Mountains

North of Copernicus you will find the Carpathian Mountains (**3**). This mountain range measures 280 by 60 km and includes many peaks over 2400 m in altitude, separated by deep valleys. You can look for the rille (**5**) that runs alongside Gay-Lussac (**4**), a 26-km-wide crater.

1. Copernicus	**5.** Gay-Lussac Rille
2. Fauth	**6.** Draper
3. Carpathian Mountains	**7.** Mare Insularum
4. Gay-Lussac	**8.** Mare Imbrium

→ A region of stark contrasts

A curious mountain range

Another region of contrasts is to fire your curiosity this evening. Mare Cognitum (**1**) stretches for 300 km between the Riphaean Mountains (**2**) to the north and Mare Nubium (**3**). The Riphaean Mountains are a strange, straight range of mountains 150 by 40 km, rising to only 1000 m.

Apollo 12 touched down 150 km north-east of these mountains on 19 November 1969.

A ghost crater and its twin

North of Mare Nubium, the almost ghost crater Lubiniezky (**4**) no longer has more than a modest 45-km ring formed by a circle of hills less than 300 m high, broken to the south-east, and surrounding a floor that is as flat and dark as the surrounding mare.

Its near twin, Kies (**5**), differs only in having a curious mountainous appendix running southwards from the crater. This 40-km feature is a marker for finding the dome Kies Pi (**6**), lying between Kies and Campanus (**7**). This 10-km diameter dome has to be viewed at low Sun angles because it is probably no more than 500 m high. A powerful telescope will show up the 2-km vent at its summit.

Bullialdus, an exemplary crater

In the centre of Mare Nubium reigns the superb 60-km-wide crater Bullialdus (**8**). Its outer slopes, furrowed by radiating valleys, stand 2000 m higher than the surrounding plain. From the crater rim, the terraced wall, with its landslips on the south-west side, plunges more than

3000 m to the narrow crater floor with a central mountain range with several 2000-m peaks. Whitish material ejected at the time of impact can be seen on the surface of Mare Nubium.

More intriguing still is the 'bridge' across a valley running out from Bullialdus. It is actually a highstanding strip of land, some 10 km wide (**9**), running alongside the valley in question west of Lubiniezky.

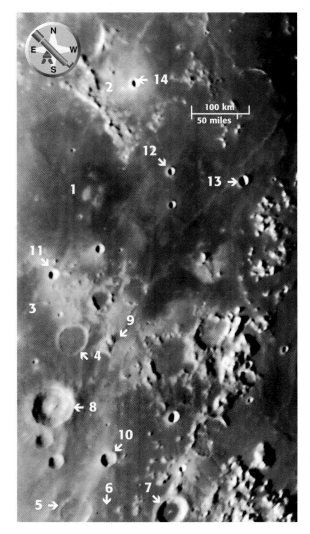

98 · Guide to the Moon ·

THE SEA THAT BECAME KNOWN

Initially the basin north of Bullialdus was considered part of Mare Nubium. But on 31 July 1964, the American Ranger 7 probe crash-landed there. Its camera took a final photo at an altitude of 1500 m just before impact. Details 40 cm large could be seen on this final snapshot, providing essential knowledge of the Moon's surface for 'soft' lunar landings. Accordingly, the International Astronomical Union decided to name this region Mare Cognitum, the Sea of Knowledge.

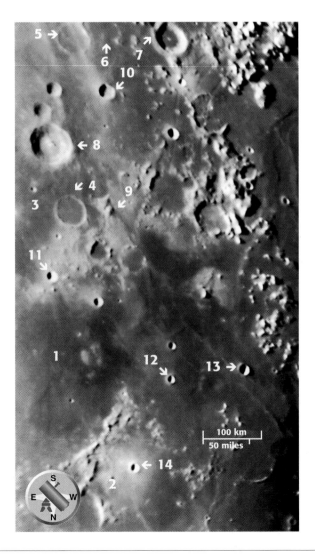

1. Mare Cognitum
2. Riphaean Mountains
3. Mare Nubium
4. Lubiniezky
5. Kies
6. Kies Pi dome
7. Campanus
8. Bullialdus
9. Bullialdus bridge
10. König
11. Darney
12. Norman
13. Herigonius
14. Euclides

→ An intensely cratered area

Clavius in majesty

A string of stunning craters awaits you this evening in the southern part of the Moon. Among them, one of the Moon's features not to be missed, Clavius (**1**), which bears the marks of its great age and is one of the largest and most complex of lunar craters.

You will immediately notice two similar craters, 45 km in diameter and 2500 m deep, which have smashed the 4500-m-high wall of Clavius: Porter (**2**), whose floor has a double central mountain and craterlets, and Rutherfurd (**3**), with a stately line of crests on its northern rim. On the floor of Clavius, which is convex because of the curvature of the Moon, notice first the arc of four craters of decreasing size. Then try to spot the exceptional alignment of three 2-km craterlets close to the south-west wall.

Longomontanus and Blancanus

North-west of Clavius, take a look around Longomontanus (**4**), an old, 145-km-wide crater with a flat floor filled with dark lava. Its terraced wall is largely ruined by many craterlets. The 3500-m-high wall rings a flat floor with a trio of off-centre hills no more than 1500 m high and containing small white patches.

On the south-west outer slopes of Clavius lies Blancanus (**5**), a crater that is clearly younger than its neighbour. It is 105 km wide with a fantastic terraced wall 4000 m high. Two craterlets and three 1000-m-high hills adorn the south-eastern part of its flat floor.

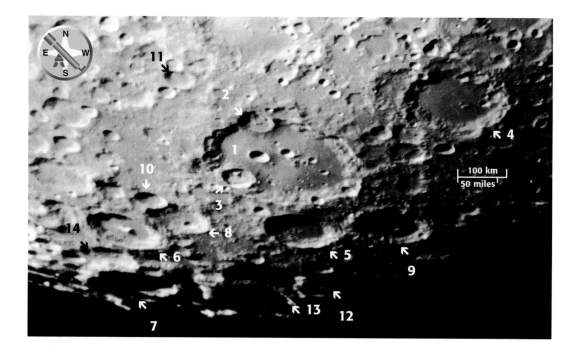

Moretus and Newton

Dwell for a while on the magnificent 115-km diameter but little-known crater Moretus (**6**). Its outer slopes rise steeply to more than 2000 m above the surrounding rough ground. Its imposing terraced walls then plunge 5000 m below the rim. The flat floor contains a few small hills and, above all, a central mountain 2700 m high: notice that its slopes do not exceed 30°.

To end this evening's excursion, have a look at Newton (**7**), an 80-km crater whose inner wall soars to 8000 m making it the deepest of the Moon's craters. A stone thrown from the crater rim would take several minutes to reach the floor. Given its position, Newton's floor is virtually never in sunlight, except during favourable librations, and so it experiences near constant night.

1. Clavius
2. Porter
3. Rutherfurd
4. Longomontanus
5. Blancanus
6. Moretus
7. Newton
8. Gruemberger
9. Scheiner
10. Cysatus
11. Deluc
12. Klaproth
13. Casatus
14. Short

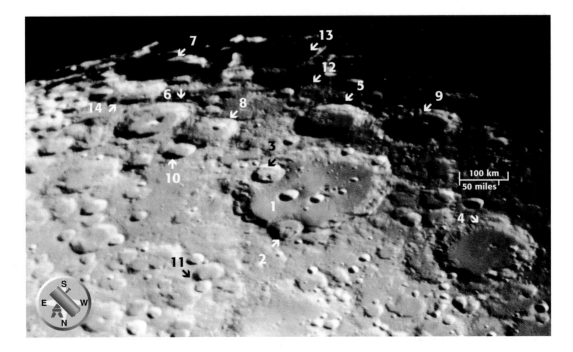

■

Evening 10

*M*idway between First Quarter and Full Moon, this evening might be called 'dome night' because you are going to discover a large number of them.

→ North of Mare Frigoris (**1**) lies John Herschel (**2**), an enormous crater 156 km across. Its immense, flat floor contains a wealth of crevasses, hills and craterlets. Its low, ruined crater wall is overlapped to the south by Horrebow, a 24-km-wide crater.

→ Mare Imbrium (**3**) can be seen in full this evening. Its circular shape, 1250 km in diameter, edged by mountains, confirms that it is really an immense impact crater. This is borne out by the presence of the first 'mascon' (see box) to be discovered beneath the 830 000 km² of its basalt plain. Mare Imbrium bears the marks of another large impact on its north-west shore. This is Sinus Iridum (**4**), edged by the Jura Mountains, which offer a landscape like no other. To the east lies the glorious Straight Range (**5**), which is unique on the near side of the Moon.

→ One of the regions of the Moon richest in domes is on show this evening. It lies south of the crater Kepler (**6**). Look for Milichius (**7**)

and its dome as well as Hortensius (**8**) and its string of six domes.

→ Further south, the day is now rising over Mare Humorum (**9**), a fantastic circular closed basin 380 km in diameter covering an area of 113 000 km². It contains a system of circular wrinkle ridges and concentric rilles, including the Hippalus Rille system (**10**) to its east, not far from the crater Agatharchides (**11**).

→ Another small basin is visible this evening: Palus Epidemiarum (**12**) with the craters Campanus and Mercator (**13**) on its northern shore, and Hesiodus Rille on its eastern edge.

1. Mare Frigoris
2. John Herschel/ Horrebow
3. Mare Imbrium
4. Sinus Iridum/Jura Mountains
5. Straight Range
6. Kepler
7. Milichius
8. Hortensius
9. Mare Humorum
10. Hippalus Rille
11. Agatharchides
12. Palus Epidemiarum
13. Campanus/Mercator
14. Hainzel

Finally, linger awhile on the curious feature Hainzel (**14**), composed of several jumbled craters in a rectangle 70 by 20 km reminiscent of Heraclitus.

Box A: *see* pp. 104–105

Box B: *see* pp. 106–107

Box C: *see* pp. 108–109

DID SOMEONE SAY 'MASCON'?

Mascon is short for 'mass concentration'. Studies of the shape of the Lunar Orbiter trajectories identified 12 regions where the internal density of the Moon is greatly increased, probably because of concentrations of mass, hence the name. Intriguingly, all these mascons are located on the near side and invariably coincide with impact basins. Some scientists believe these mascons are the remains of the huge meteorites that created the maria and remain buried at depth.

→ The Bay of Rainbows

Sinus Iridum

Sinus Iridum (**1**) is without contest the most beautiful feature of its kind. This 400-km-wide half-crater extends over a total area of 237 000 km^2 and is thought to have formed before Mare Imbrium (**2**) and subsequently filled with lava from the basin. This theory is supported by the fact that the northern edge of the bay lies 600 m below 'sea level', as if the lava was unable to fill it completely.

The Jura Mountains

The Jura Mountains (**3**) are vestiges of the outer slopes of the original crater. They are composed of several lines of parallel, concentric mountains 30 km wide with peaks soaring to altitudes of 4000 m. Three of them

THE MOON MAIDEN

The Heraclides Promontory on the edge of Sinus Iridum takes on a strange appearance when the Moon is 10½ days old. A woman's head appears, sometimes young, sometimes ugly with a jutting chin . . . and more often than not a simple, somewhat indistinct caricature. The variable appearance of this feature depends on the conjunction of libration and phase of the Moon at the moment of observation. Cassini was the first to report it and marked it on his 1692 map of the Moon. In his admirable photographic atlas published in 1986, Georges Viscardy pointed out that one of the buttresses of the Jura Mountains looked like a human face (the Stone Face, **4**).

resemble a stone face (**4**) when the light falls on them at a certain angle.

The eastern end of the Jura Mountains is known as Laplace Promontory (**5**). This 2500-m craggy headland drops steeply away into Mare Imbrium. To the west, Heraclides Promontory (**6**) is less impressive than its opposite number, rising to just 1700 m in a series of terraces.

The Jura Mountains were smashed almost in their centre by Bianchini (**7**), a 40-km-wide crater some 3000 m deep whose faintly terraced wall surrounds a flat floor with a small hill set off from its centre.

Another dramatic formation that can be seen tonight is the Straight Range (**8**), also known as Montes Recti. It is a straight mountain massif 100 km long and 20 km wide rising to 1800 m and with a craterlet 50 km wide near its western end.

1. Sinus Iridum	**6.** Heraclides Promontory	**11.** Helicon
2. Mare Imbrium	**7.** Bianchini	**12.** Le Verrier
3. Jura Mountains	**8.** Straight Range	**13.** Teneriffe Mountains
4. Stone Face	**9.** Maupertuis	
5. Laplace Promontory	**10.** Condamine	

→ The lunar volcano range

Kepler and its surroundings

Surrounded by a plateau of rugged relief as testimony to the violence of the impact that created it, Kepler (**1**) is 2800 m deep and 35 km in diameter. Its steep ejecta blanket rises 1000 m above the surrounding plateau and gives on to a slightly terraced crater wall. The rough crater floor has only a slight central elevation and a few hummocks and craterlets. East of Kepler, look for Milichius (**2**), a bowl-like crater 13 km in diameter and 2000 m deep. Beside it, Milichius Pi (**3**) is an incredible isolated volcanic dome 10 km across and capped by a 2-km vent which can be seen with a powerful telescope.

South of the crater Tobias Mayer (**4**), the low Sun angle along the terminator also reveals other unnamed domes (**5**), some of which have small summit pits. These domes are lower than that of Milichius and therefore more difficult to spot.

Hortensius and the range of domes

Now look 200 km south of Milichius for Hortensius (**6**), a bowl-shaped crater barely larger than Milichius. There, not one but six domes await your visit (**7**)! They lie in an arc, perhaps marking an underlying fault along which

material rose from the Moon's interior. They range from 7 to 13 km in diameter. Of these six domes only one has no vent, four have vents, and the other has a double vent. Look for them with a powerful telescope under oblique lighting. These volcanoes probably became extinct several million years ago as no transient phenomena have been detected in their immediate vicinity.

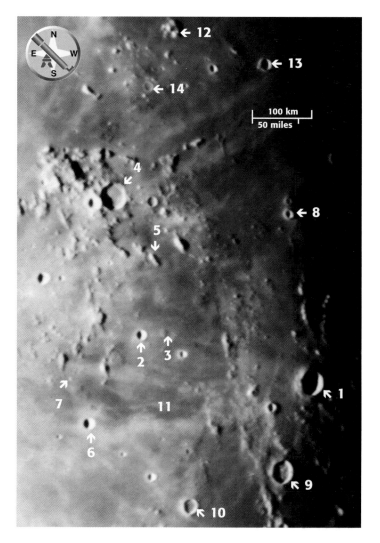

THE MOON AND FILTERS

A neutral density 'Moon' filter of the type often supplied with a telescope may be screwed to the front of the eyepiece to reduce the Moon's glare. These filters are only useful at low magnifications in telescopes with apertures of more than 120 mm. It is best to choose a polarising filter so that glare can be adjusted as required by turning it in its mount. To study the details of lunar features a blue no. 80 A or yellow no. 21 filter heightens contrasts.

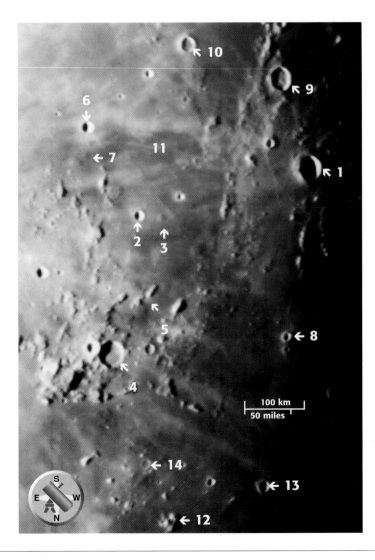

1. Kepler
2. Milichius
3. Milichius Pi
4. Tobias Mayer
5. Tobias Mayer domes
6. Hortensius
7. Hortensius domes
8. Bessarion
9. Encke
10. Kunowsky
11. Mare Insularum
12. Mount Vinogradov
13. Brayley
14. Natasha

➔ The Marsh of Epidemics

Palus Epidemiarum

On the southern edge of Mare Nubium (**1**), the fluid lava overflowed and invaded a low-lying region giving rise to Palus Epidemiarum (**2**) or Marsh of Epidemics, which is easily recognisable from the two craters bordering it, Campanus (**3**) and Mercator (**4**).

Campanus has a flat floor, 50 km across, filled with dark lava, and contains a low central peak and a few craterlets, one of which is easy to spot. The 2000-m-high terraced wall has steep outer slopes on the Mare Nubium side.

Mercator is a near twin of Campanus, although not quite as wide. The craterlet Mercator B overlaps the 1800-m-high eastern wall. The wall extends southwards as a remarkable S-shaped mountain crest 50 km long. The dark, flat crater floor bears no features that can be seen with an amateur telescope.

A festival of rilles

The two craters Campanus and Mercator are separated by the fine Campanus Rille (**5**) and differ only in that Campanus has a central hill. They are extended to the south-east by a 180-km-long scarp known as Rupes Mercator (**6**).

Spend a moment on Capuanus (**7**), an old, ruined crater some 60 km in diameter with an asymmetric wall. The wall rises to nearly 3000 m on the western side

but is barely more than 500 m high to the north-east. The crater floor is flooded with dark, fluid lava and has another point of interest – it has volcanic domes although they are difficult to make out.

Palus Epidemiarum contains the fine straight Hesiodus Rille (**8**). This cleft is 300 km long but only 3 km wide and crosses

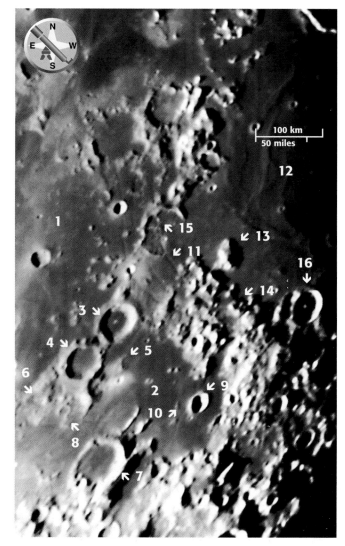

several formations, including hills, and features several craterlets that formed subsequently.

Look next between Ramsden (**9**) and Capuanus (**7**) for a network of branching rilles extending over 130 km and looking similar to those of Triesnecker. These are the Ramsden Rilles (**10**).

Some of them branch at right angles, making it difficult to come up with a theory about their origin.

As the highlight of this outing do not miss the Hippalus Rilles (**11**). These are three wonderful parallel, concentric wrinkles about 240 km long which can be seen north-west of Campanus circling Mare Humorum (**12**) at some distance. Their shape suggests they are associated with the formation of this basin when the fluid lava went through a cooling phase.

To finish off, have a look at Kelvin Promontory (**13**), a lone mountainous cape 35 by 25 km close to the 150-km-wide Rupes Kelvin (**14**) that forms the south-eastern rim of Mare Humorum.

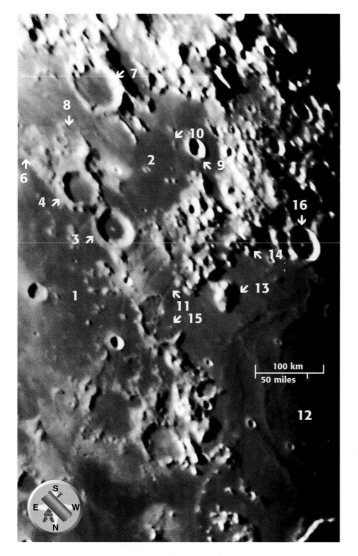

1. Mare Nubium
2. Palus Epidemiarum
3. Campanus
4. Mercator
5. Campanus Rille
6. Rupes Mercator
7. Capuanus
8. Hesiodus Rille
9. Ramsden
10. Ramsden Rilles
11. Hippalus Rilles
12. Mare Humorum
13. Kelvin Promontory
14. Rupes Kelvin
15. Hippalus
16. Vitello

Evening 11

*T*he Moon is growing rounder and this evening's terminator reveals more varied features than it does craters.

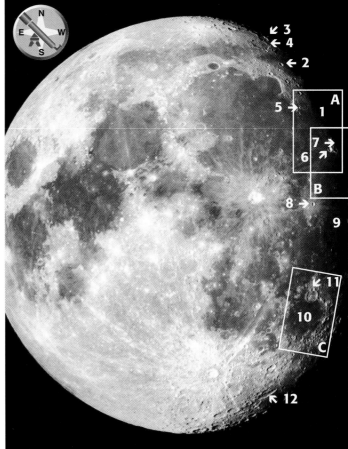

→ In the north, Mare Frigoris is extended westward by Sinus Roris (**1**). On the boundary between the two, the attractive 40-km-wide crater Harpalus (**2**) is a miniature replica of Copernicus, with its 2800-m terraced wall and triple central peak.

→ Above Mare Frigoris is the old, 70-km-wide, ruined crater Anaximander (**3**). It lies near the very fine John Herschel (**4**) with its 150-km diameter flat floor and low ramparts.

→ Do not miss out on observing Mounts Gruithuisen Delta and Gamma (**5**) at the end of the mountain range extending the Jura Mountains westwards. Gruithuisen Gamma is an extraordinary dome to be studied in priority. Not far away, Prinz and its rilles will also catch your eye. Discover too the very interesting region of Aristarchus (**6**) with the famous Schröter's Valley (**7**) and the crater Herodotus.

→ See how Kepler (**8**) now reveals the cobweb around it with some of its rays joining up with those of Copernicus. All around it, the Sun is rising a little more every day over Oceanus Procellarum (**9**).

→ Mare Humorum (**10**) offers us this evening's other star features with the scenic crater Gassendi (**11**), Rupes Liebig and the region of Mersenius.

→ Lastly, do not miss the oddity Schiller (**12**), the only crater on the Moon of very elongate shape. This feature is 180 km long but only 70 km wide and seems to have resulted from

1. Sinus Roris
2. Harpalus
3. Anaximander
4. John Herschel
5. Gruithuisen Delta and Gamma
6. Aristarchus
7. Schröter's Valley
8. Kepler
9. Oceanus Procellarum
10. Mare Humorum
11. Gassendi
12. Schiller

Box A : *see* pp. 112–113

Box B : *see* pp. 114–115

Box C : *see* pp. 116–117

the merger of at least two craters, unless it was produced by a very low angle meteorite impact. The 4000-m-high walls surround a floor that is clearly flat in the south and pitted by a few craterlets. In the north, a small central mountain rises together with a few hills that contrast with the southern part of the crater floor.

HOW TO MEASURE A LUNAR CRATER

Measuring the diameter of lunar craters is within the capability of any amateur astronomer with a telescope, a reticule eyepiece and a timepiece. Stop the clock-drive of the mount and measure the time it takes for the Moon's equator to move across a vertical marking. Given the mean diameter of the Moon (3500 km), you can deduce its speed as x km/s. You then measure the time it takes for the crater to move across the vertical marking. Multiply this time, in seconds, by x to give the crater's diameter.

→ Islands in the storms

The Gruithuisen Mountains

Beyond the Heraclides Promontory, the mountains forming the rampart to Mare Imbrium drop away westwards into Oceanus Procellarum (**1**) by way of a number of glorious formations, leaving only a few islands emerged.

First are Mounts Gruithuisen Gamma (**2**) and Gruithuisen Delta (**3**), two similar-looking characteristic features but which are probably of different origins, located north of the crater Gruithuisen (**4**).

Gruithuisen Gamma is an enormous circular dome 20 km across and rising to probably 1000 m. It can be seen even at high

Sun angles. Its flattish summit has a 2-km diameter vent which is visible in a 150-mm telescope.

Alongside it, to the south-east, Mount Gruithuisen Gamma looks more like a straightforward mountain as there is no vent to be seen at its top. While similar in size to its neighbour, it may be no more than a relic of the ramparts of Mare Imbrium.

Harbinger and Prinz

Continue your trip 300 km southwards, where you come upon the Harbinger Mountains (**5**), an isolated group of three main peaks

THE MAN WHO SAW A CITY

When observing the Moon, do not let your imagination run away with you. That is just what happened in the nineteenth century to Baron von Gruithuisen, who wrote a book about constructions erected by a 'Selenite' civilisation he claimed to have seen on the Moon. Of course, what he had seen were peculiar landforms but the poor performances of his telescope had led him astray. Not everyone learned from the lesson, as it recurred in the 1980s with the business of the 'face' on Mars recorded by the Viking probes, which is really just a plateau whose unusual shape was finally revealed by the higher-resolution cameras of Mars Global Surveyor.

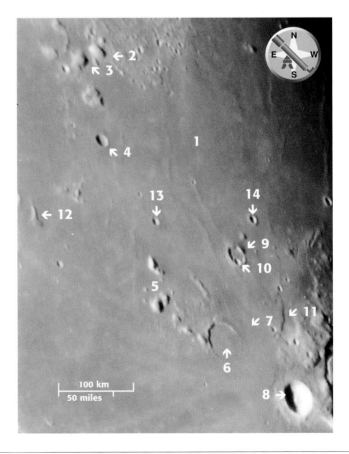

stretching over some 90 km. They rise up to 2500 m from the lava of Mare Imbrium and are further remains of its ancient walls. Its peaks, illuminated by the rising Sun, herald sunrise over Aristarchus.

The Harbinger Mountains stand close to the entrancing semi-ghost crater Prinz (**6**). Its 50-km diameter wall remains only in the north-east where it struggles to reach 1000 m in height.

North of Prinz, entwined in the foothills of the Harbinger Mountains, try to spot the fine rilles (**7**), some of which run as far as Aristarchus (**8**). The most conspicuous one zigzags from the northern edge of Prinz towards Krieger (**9**).

Krieger – a half-engulfed crater

Krieger is a half-drowned crater whose walls rise only 1000 m above the surrounding lava. The crater wall was smashed in the south by Van Biesbroeck (**10**), a 10-km craterlet some 1000 m deep, its floor being even deeper than Krieger's own floor. Notice, to finish, Rupes Toscanelli (**11**), a 70-km-long fault, not as well known as the Straight Wall.

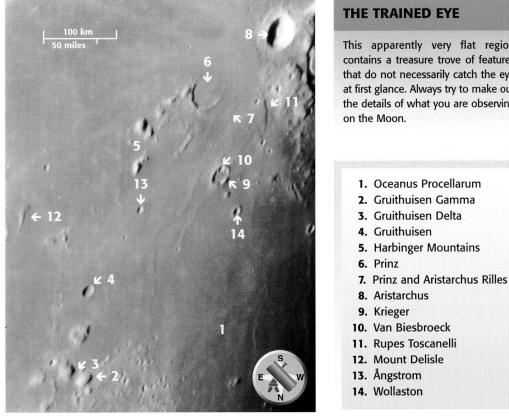

1. Oceanus Procellarum
2. Gruithuisen Gamma
3. Gruithuisen Delta
4. Gruithuisen
5. Harbinger Mountains
6. Prinz
7. Prinz and Aristarchus Rilles
8. Aristarchus
9. Krieger
10. Van Biesbroeck
11. Rupes Toscanelli
12. Mount Delisle
13. Ångstrom
14. Wollaston

→ The lunar lighthouse

Aristarchus – the brightest of the craters

Watch out, here is one of the most famous regions of the Moon because this is where the vast majority of lunar transient phenomena have been reported. It is true too that this geologically very interesting region seems conducive to them.

Let us begin with the superb crater Aristarchus (**1**). Despite its modest 40-km diameter, it is the brightest of the Moon's craters and a real searchlight. It can even be seen by the earthshine. A long tongue of the underlying white material has also been projected towards Herodotus (**2**) and some of the rays extend as far as Kepler and even Copernicus.

Aristarchus is thought to be 450 million years old, between the ages of Copernicus and Tycho. Aristarchus is 3000 m deep, with terraced walls featuring a few dark slumps surrounding a narrow, flat floor with a dazzling central peak standing no more than 500 m high.

Herodotus is very different from its neighbour Aristarchus and is older too, its lava-filled floor indicating that it was formed before Oceanus Procellarum (**3**). Its 35-km diameter wall is no more than 1500 m high and has been smashed by a craterlet in the north.

North of Aristarchus you

can admire a remarkable, rugged plateau, which is probably also of volcanic origin and which terminates to the east in Rupes Toscanelli (**4**). In the north, this plateau joins up with the scenic straight range of the Agricola Mountains (**5**), which have a very characteristic appearance, being 160 km long, narrow at only 10 km wide, and no more than 1500 m high.

The finest lunar rille

On the northern outer slopes of Herodotus

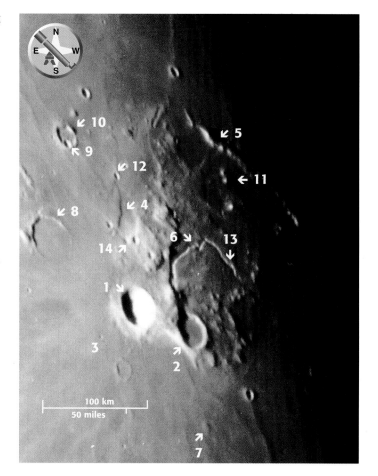

you cannot miss Schröter's Valley (**6**). This is the easiest rille to spot anywhere on the Moon. It runs for 160 km northwards from Herodotus and then slants westward. It varies in width from 6 to 10 km, narrowing to about 500 m at its western end. The valley starts from an elongate craterlet forming the 'Cobra's Head', which is probably a volcanic vent. The valley itself is thought to be an ancient lava tube with a collapsed roof, similar to those found on Earth on the sides of Hawaiian volcanoes . . . only much larger!

Some 70 km south of Herodotus you will also notice the small dome Herodotus Omega (**7**), to be observed at low Sun angles.

KEEP AN EYE ON ARISTARCHUS!

The catalogue of lunar transient phenomena kept by NASA shows that Aristarchus is one of the most commonly listed craters for 'unusual' sightings. Keep an eye on this apparently still geologically active region, particularly as it bears all the hallmarks of a volcanic domain.

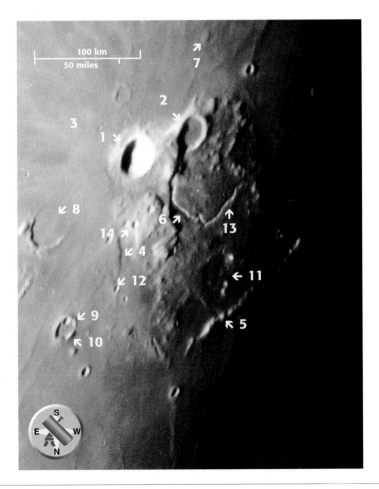

1. Aristarchus
2. Herodotus
4. Rupes Toscanelli
3. Oceanus Procellarum
5. Agricola Mountains
6. Schröter's Valley
7. Herodotus Omega
8. Prinz
9. Van Biesbroeck
10. Krieger
11. Herodotus Mountains
12. Toscanelli
13. Freud
14. Vaisälä

→ Mare Humorum

A circular basin

Almost a twin of Mare Crisium, Mare Humorum (**1**) awaits your visit. This fine example of a circular basin has interesting landforms, relics of its origins. At 380 km in diameter for an area of 113 000 km², it contains a system of concentric wrinkle ridges (**2**) in its eastern part.

Then, whereas to the east you were lucky enough to see the Hippalus Rilles (**3**) yesterday, to the west today you can admire the two parallel Mersenius Rilles (**4**), some 230 km long. Close by you can also view Rupes Liebig (**5**), an 80-km-long fault extended northward by a rille.

The lava of Mare Humorum has invaded several ancient craters. First look for Doppelmayer (**6**), a 65-km crater whose wall is completely buried to the north, but not so the 2000-m central mountain.

East of Doppelmayer is the 25-km-wide ghost ring of Puiseux (**7**) whose walls are no more than 400 m high. Then, south of Doppelmayer, you will see a second, half-immersed, 40-km-wide crater Lee (**8**).

Gassendi, the pearl ring

But the high point of this evening's display lies on the northern edge of Mare Humorum.

Gassendi (**9**) is a glorious crater not unlike Posidonius. Together with the crater Gassendi A, excavated in its northern wall, it looks like a ring adorned with a pearl. This splendid walled plain 90 km wide is surrounded by a wall that does not exceed 2000 m to the north-west. The floor of Gassendi is home to an impressive network of rilles surrounding a triple central mountain 1200 m high and many hills.

Reserve a brief visit for Mersenius (**10**), an 80-km-wide crater some 2500 m deep, with a curious convex floor adorned with a chain of craterlets.

Finally, south of Mersenius between De Gasparis (**11**) and Liebig (**12**), you will observe the De Gasparis Rilles.

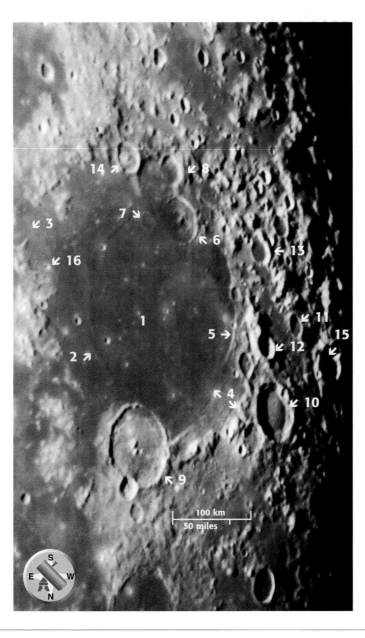

1. Mare Humorum
2. Wrinkle ridge system
3. Hippalus Rilles
4. Mersenius Rilles
5. Rupes Liebig
6. Doppelmayer
7. Puiseux
8. Lee
9. Gassendi
10. Mersenius
11. De Gasparis
12. Liebig
13. Palmieri
14. Vitello
15. Cavendish
16. Loewy

Evening 12

The Moon becomes even rounder and offers some real curiosities this evening. But there are fewer of them because it becomes more difficult to observe the smaller landforms owing to the angle of sight from Earth.

➜ Concentrate first on the extraordinary Mount Rümker (**1**) located at the extreme south of Sinus Roris (**2**). This feature is worth a detour because it is the only complex of clustered domes on the near side. They form a rugged, circular plateau some 80 km across but no more than 500 m high.

➜ Some 500 km further south, near the crater Marius (**3**), you will come upon the largest field of separate domes of the near side.

➜ Just north of the equator, Reiner (**4**), a 30-km-wide crater, stands out prominently. It will guide you to Reiner Gamma (**5**), a kite-shaped white patch of mysterious origin. Another 200 km to the west of this patch, the Soviet probe Luna 9 was the first to touch down, on 3 February 1966.

➜ South of the equator, a huge, dark circle will draw your eye. This is Grimaldi (**6**), which, with the enormous Riccioli and the interesting Hevelius, forms a truly gigantic threesome.

→ Further south, the dark circle of Billy (**7**) and the bright ring of Hansteen are a pair of remarkable 45-km diameter craters of totally contrasting appearance.

→ Tonight you will also be able to admire the longest lunar cleft, Sirsalis Rille (**8**). It snakes between many craters and ends in the south near to Lamarck and Darwin (**9**), which are very much ruined. Crüger (**10**) and Zupus (**11**) will draw your attention too as their darkness contrasts with the landscape.

→ To end your excursion, a superb trio awaits you in the south-west of the disc. This is the group including Schickard (**12**), the strange Wargentin with its raised floor, and the old Phocylides.

Box A: *see* pp. 120–121
Box B: *see* pp. 122–123
Box C: *see* pp. 124–125
Box D: *see* pp. 126–127

1.	Mount Rümker	**7.**	Billy
2.	Sinus Roris	**8.**	Sirsalis Rille
3.	Marius	**9.**	Lamarck
4.	Reiner	**10.**	Crüger
5.	Reiner Gamma	**11.**	Zupus
6.	Grimaldi	**12.**	Schickard

→ A field of domes

Marius and its region

This evening you are to visit one of the strangest regions of the Moon, located just south of Aristarchus (**1**) beside the crater Marius (**2**).

Marius itself is interesting: its 40-km-wide floor was filled by runny lava from Oceanus Procellarum (**3**). Accordingly, the ramparts now rise a mere 1500 m above the surface. Then two craterlets formed, one on the south-east outer slope and the other a 3-km crater pit in the crater floor. It is an excellent test for 100–120-mm telescopes . . .

But it is north and west of Marius that you really need to look. The terrain looks lumpy, stippled with domes no more than 1000 m high. This area is well delimited. It is therefore a region that experienced a particular type of volcanic activity. This theory is reinforced by the somewhat indistinct presence of a sinuous rille,

250 km long and 2 km wide, north of the dome field (**4**).

Some of the domes have vents at their tops, no more than 2 km wide, but it takes a 200-mm telescope to pick them out. In fact, the Marius region viewed from the Moon's surface must look something like the volcanoes of the Auvergne region of central France.

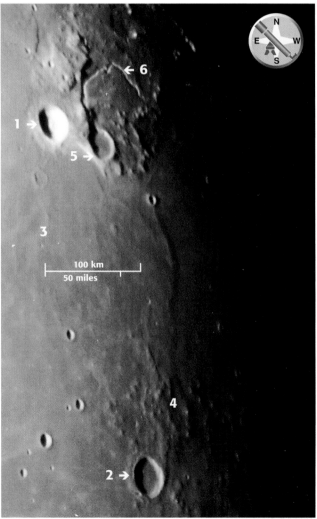

IS THE MOON A DEAD WORLD?

The Moon is virtually a dead world today. Seismometers set up by the Apollo missions have detected 'moonquakes' that do not exceed 2 on the Richter scale. The solidified crust is some 60 km thick and below it the solid mantle extends to a depth of about 800 km. It is likely that there is still a molten core 1000 km in diameter at the centre of the Moon.

THE MOON ON COMPUTER

Those lucky enough to own a computer will probably be frustrated to learn there are few software packages about our satellite. Even so we can mention two pieces of freeware: Lunarcalc by the Canadian Alister Ling and Virtual Moon by Christian Legrand and Patrick Chevalley. As for atlases on CD-ROM, there are only those by Lunar Orbiter and *Clementine*.

An unelucidated mystery

Because of its geological interest this area might have been visited by astronauts had the Apollo missions not been discontinued. Were there enormous gas bubbles that rose as the lava solidified or upwellings of underlying magma that lifted the already cold crust? Only analysis of samples will reveal the secrets of the origins of this part of the Moon.

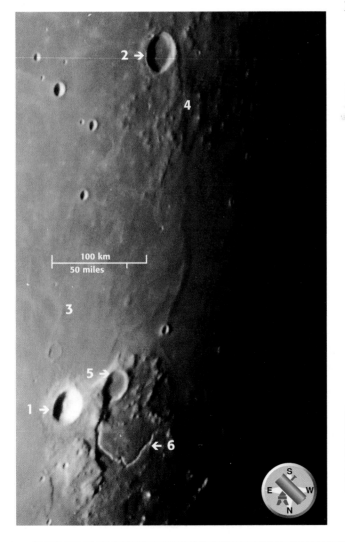

1. Aristarchus
2. Marius
3. Oceanus Procellarum
4. Dome field
5. Herodotus
6. Schröter's Valley

→ A trio of giants

Grimaldi

On the western limb of the Full Moon, Grimaldi (**1**) is a 220-km-wide feature which looks like a scaled-down Mare Crisium. Observed through a telescope, its position is an excellent indicator of the direction and intensity of the libration.

Its flat floor is filled with lava that is darker still than that of the nearby Oceanus Procellarum but which is of unknown origin. Its very low wall blends into the landscape, except in the south-east where a 3000-m mountain casts a giant shadow.

Riccioli

North-west of Grimaldi another huge crater is coming into view. This is Riccioli (**2**) whose 150-km diameter wall, higher in the east than the west, is far more conspicuous than that of its neighbour. The floor of Riccioli is very rugged with a wealth of hummocks and craterlets. A network of rilles that are hard to spot scars the floor, which has an enigmatic dark patch in its northern part.

If you have at least a 200-mm telescope, look for the many rilles that scour this region. Grimaldi Rille (**3**) starts from Damoiseau (**4**) and skirts the south-east wall of Grimaldi.

Hevelius

Finally, north of Grimaldi, and separated from it by Lohrmann (**5**), lies Hevelius or Hevel (**6**), another walled plain 110 km in diameter whose flat floor also features a network of rilles ringing a small, central mountain.

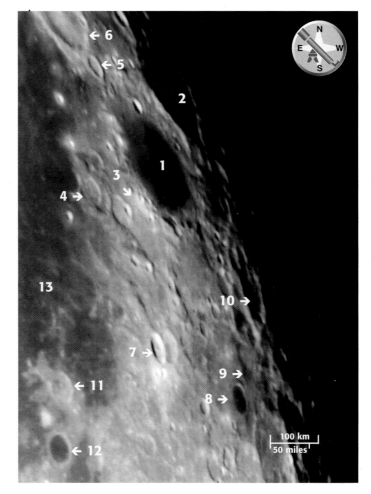

THE FATHERS OF THE LUNAR NAMING SYSTEM

Grimaldi and Riccioli, two Jesuits priests, are famous for initiating the naming system of lunar landforms still used today. Riccioli published the *Almagestun Novum* in 1651 with a map of the Moon engraved by Grimaldi. To name the formations Riccioli used only the names of ancient or deceased philosophers and scientists, without forgetting his own name and that of Grimaldi. In this way he avoided the criticism that befell his predecessor Hevelius, who, on his 1647 map, had used the names of celebrities of the day. Those left out must not have appreciated it! Since that time, the naming system has remained largely unchanged except for regular additions voted by the International Astronomical Union, which now manages the system.

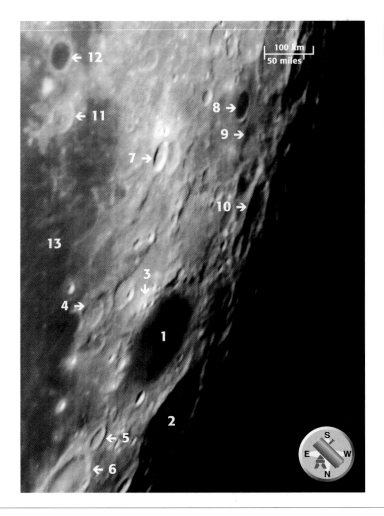

1. Grimaldi
2. Riccioli
3. Grimaldi Rille
4. Damoiseau
5. Lohrmann
6. Hevelius
7. Sirsalis
8. Crüger
9. Lacus Aestatis
10. Rocca
11. Hansteen
12. Billy
13. Oceanus Procellarum

→ The longest lunar rille

Sirsalis Rille

The rille-rich region south of Grimaldi features the longest rille on the near side. To find it you first need to pick out the pair of craters Sirsalis/Sirsalis A (**1**), each some 40 km in diameter.

Sirsalis is younger than Sirsalis A as it overlaps the eastern rampart of the latter and its white ejecta are still visible. Sirsalis also differs from its elder companion in having a more distinctly terraced wall and a central mountain.

Sirsalis Rille (**2**) runs for nearly 400 km from the crater Sirsalis, across the crater Byrgius A (**3**), which is older than the rille, and ends near Byrgius (**4**). It is probably a graben 2–3 km wide that appeared when the lunar crust cracked at the time of some major event. The proximity of Mare Orientalis may be an explanation.

Two interesting craters

The region is also home to the strange crater Crüger (**5**), the twin of Billy. This 45-km-wide crater has low walls rising to between 500 and

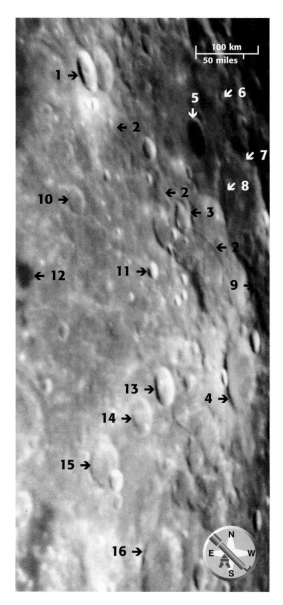

1. Sirsalis/Sirsalis A
2. Sirsalis Rille
3. Byrgius A
4. Byrgius
5. Crüger
6. Lacus Aestatis
7. Darwin
8. Darwin Rilles
9. Lamarck
10. Fontana
11. De Vico
12. Zupus
13. Henry Frères
14. Henry
15. Cavendish
16. Vieta

ARE CRATERS STILL BEING FORMED?

An event in the year 1178 provides something of an answer to this question. Monks at Canterbury reported seeing a strange phenomenon on the Moon's surface. A 'flaming torch' supposedly appeared at the top right of the lunar crescent. The Apollo missions proved that at this location, on the boundary with the far side, there is very young 15-km crater, since named Giordano Bruno. But nothing proves that the two are related . . .

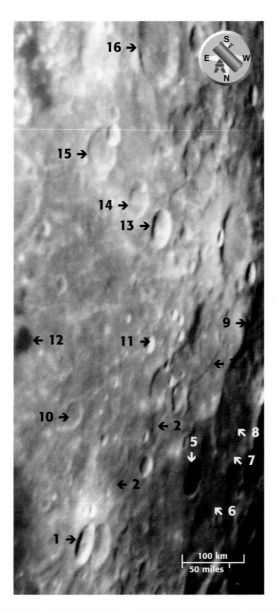

1000 m around a lake of very dark lava devoid of any craterlets. But unlike Billy, which lies on the edge of a lava-producing basin, Crüger is in the midst of the highlands. The lava of its floor must necessarily then have risen locally from deep below the surface.

Byrgius, at the southern end of the Sirsalis Rille, is a remarkable crater, 90 km wide. Its wall is more than 3000 m high in places and has been smashed to the east by a 20-km diameter crater. This impact threw up bright material, sprays of which settled over the flat floor of Byrgius.

→ Three definitely not of a kind

The very large crater Schickard

Figuring among the largest of the Moon's craters at more than 230 km in diameter, Schickard (**1**) is a huge enclosure surrounded by walls rising to 1500 m on average but with a few peaks of more than 2000 m.

The floor of Schickard is pocked with craterlets and has a number of isolated hills. In the south-western part, gigantic parallel gouges appear to have been cut by the impact that formed Schiller. Stranger still is the mosaic of dark and light cobbles that strew the floor of this exquisite crater. Could this be volcanic fallout similar to that in Alphonsus?

Wargentin, the overbrimming crater

South-west of Schickard lies one of the strangest of the Moon's craters, Wargentin (**2**). As you can see, the floor of Wargentin has filled with dark lava as with many other craters. But the unique feature here is that the

lava continued to ooze out until it filled the crater to the brim, transforming Wargentin into a circular plateau 85 km in diameter and some 400 m high.

This is probably because the lava did not

WHAT IS REGOLITH?

The Moon's surface is entirely covered in dust to depths ranging from a few millimetres to several metres depending on location. This is the regolith. Its texture is reminiscent of wet sand, it is dark coloured and it sticks fast to clothing. But the most surprising thing about it is the odour the astronauts noticed when they removed their soiled spacesuits back inside the lunar module – it was like gunpowder! The regolith comes from a number of sources: the constant showering of micrometeorites, solar radiation, the amplitude of temperature fluctuation between night and day, and various modes of erosion all eventually wear down the surface rocks to powder.

find any breach in the wall through which to spill out over the surrounding terrain. Then, when it cooled, this lava lake creased to form the Y-shape that is conspicuous on its surface.

The old Phocylides

Connected to Wargentin by the old ruined crater Nasmyth (**3**), Phocylides (**4**) is the final member of the trio and is another walled plain 115 km in diameter. Its terraced wall was smashed by Phocylides F in the south and has slumped to the north. Its flat floor has many craterlets, the largest of which is in the south-west, together with small hills.

1. Schickard
2. Wargentin
3. Nasmyth
4. Phocylides
5. Drebbel
6. Nöggerath

Evening 13

*T*his is the final evening when a slight relief remains visible along the western edge of the disc. The terminator shows profile views of craters. It is a highlight when landscapes are revealed as if you were flying over them at low altitude, particularly if you turn your telescope tube so they are seen 'flat on'.

→ Perched near the northern end of the terminator, the most noticeable crater is Pythagoras (**1**). It yields a majestic view of its 130-km diameter walls, its 5000-m-high terraces and its double central mountain piercing its lava-filled flat floor.

→ Further south, Oceanus Procellarum (**2**) wraps just round to the far side. The bay of the crater Eddington (**3**) is easily visible at

135 km in diameter. The floor of this crater has been filled by lava from the neighbouring basin that breached its southern rampart. Beside it, do not miss Seleucus (**4**), a pretty, 43-km diameter crater with a terraced wall 2300 m high and a small central mountain.

→ Just below Eddington the crater pair Cardanus/Krafft (**5**), each 50 km wide, has a

AND LUNAR ECLIPSES?

When the Full Moon moves into the Earth's shadow there is a lunar eclipse, which turns it a lovely reddish hue for several tens of minutes. Dates of the next eclipses:

16 May 2003
9 November 2003
4 May 2004
28 October 2004
24 April 2005
17 October 2005
14 March 2006
7 September 2006
3 March 2007
21 February 2008
16 August 2008

No, honestly. It is possible to see the edge of the far side of the Moon thanks to its librations. A strip 200 km wide comes temporarily into view as the librations vary, revealing in particular Mare Humboldtianum, Mare Smythii, Mare Australe and Mare Orientale along with numerous craters. Likewise, it becomes possible to see beyond the Moon's north and south poles.

remarkable 60-km-long rille joining the two craters.

➤ Further south again, use the dark rings of Riccioli (**6**), Grimaldi (**7**) and Schickard (**8**) to try to find Mare Orientalis, the gigantic impact basin located mostly on the far side. This basin often only reveals its eastern walls, which are the Cordillera Mountains (**9**).

➤ Finally, this evening is the best time to observe the largest crater on the near side, Bailly (**10**). But although this monstrous feature measures 300 km in diameter, its walls stand out little from the neighbouring craters and its unfavourable position makes it difficult to observe.

Box A: *see* pp. 130–131

1. Pythagoras
2. Oceanus Procellarum
3. Eddington
4. Seleucus
5. Cardanus/ Krafft
6. Riccioli
7. Grimaldi
8. Schickard
9. Cordillera Mountains
10. Bailly

→ Mare Orientale and the Cordillera Mountains

A recent basin

The last lunar basin to have formed is known as Mare Orientale. Since 1961, when the International Astronomical Union changed the cardinal points of the Moon, this 'Eastern Sea' has found itself on the *western* limb.

This basin is more than 300 km in diameter and, as it spans the boundary between the near and far sides, only the eastern part of it can be seen from Earth. In fact, we can see short arcs of the two concentric rings of mountains that surround it and, when there are extremely favourable librations, the edge of the dark floor of the basin itself.

The Cordillera Mountains

The first of these rings is the Cordillera Mountains (**1**). Given its 900-km diameter, this mountain range appears to form a straight line. Many of its peaks soar to more than 5000 m.

To pick them out, use the string of large craters and dark-floored craters revealed by

A FINAL INFERNO

Mare Orientale is the youngest of the great impact craters that mark our satellite's surface. This infernal event must have occurred 3.8 billion years ago, and since then few new craters have marked this feature, leaving it much as it must have looked initially. It is thought that a 70-km meteorite struck the Moon at this spot at more than 40 000 km/h!

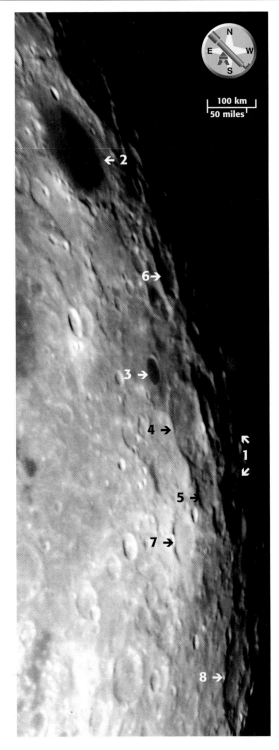

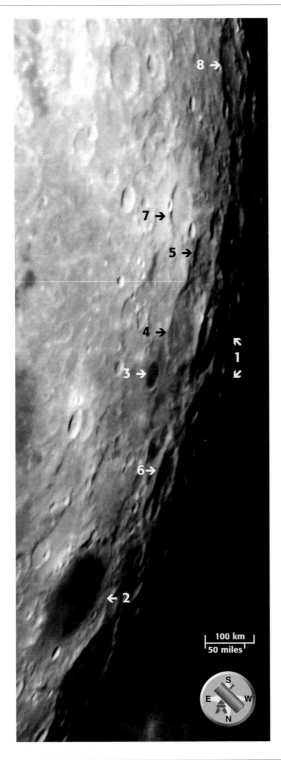

this evening's terminator. Starting from Grimaldi (**2**), head for Crüger (**3**), Darwin (**4**) and Lamarck (**5**). The Cordillera Mountains lie behind the latter craters.

The Rook Mountains

The second of these mountain rings is known as the Rook Mountains and only comes into sight when the librations are favourable. The range is 600 km in diameter and again altitudes come close to 6000 m. The 150-km-wide region separating the Cordillera and Rook Mountains is extremely rugged, evidence of the violence of the impact to which these mountains bear witness. Try to spot the dark patch of Lacus Veritatis at the latitude of Crüger.

1. Cordillera Mountains
2. Grimaldi
3. Crüger
4. Darwin
5. Lamarck
6. Rocca
7. Byrgius
8. Lagrange

Evening 14

Your calendar shows that it is Full Moon in a few hours' time. Our satellite rises at sunset and you will now have to wait a few hours until it is well clear of the horizon.

➜ The Moon is now behind the Earth, either above or below its shadow. The Sun's rays fall vertically on the centre of the disc and at low angles around its edges. Accordingly the relief is no longer brought out by giant shadows.

So what should you look for? First notice the intense luminosity of the Full Moon. So as not to be dazzled, you will have to fit a lunar filter or adjustable polarising filter, or magnify greatly to reduce the glare.

➜ Full Moon is the ideal time for discovering young craters recognisable by the bright ejecta around them. The most spectacular ejecta rays are those of Tycho (**1**), some of which reach right across the face of the Moon, along with those of Copernicus (**2**) and Kepler (**3**).

➜ There are also a very large number of bright craters but without ejecta: Langrenus (**4**), Proclus (**5**), Dawes (**6**), Dionysius (**7**), Manilius (**8**), Bode (**9**), Mösting A (**10**), Birt (**11**), Pytheas (**12**), Gambart A (**13**), Euclides

COLOURS ON THE MOON

At Full Moon with low magnification you will sometimes see delicate colours. They range from white to yellowish via various shades of grey. You can occasionally see subtle greenish hues in Mare Crisium, Mare Humorum and Mare Serenitatis. Photos confirm these faint tints, but to the untrained eye the Moon often appears just a pale yellow.

(**14**) and, brighter still, Aristarchus (**15**). But you will easily find other bright craters for yourself.

→ Have a go at spotting the large craters you have looked at over each previous evening, such as Posidonius (**16**), Aristoteles and Eudoxus (**17**), Archimedes (**18**) or Gassendi (**19**). You will be surprised how they look when illuminated vertically. Notice that the famous trios Theophilus–Cyrillus–Catharina (**20**) and Ptolemaeus–Alphonsus–Arzachel (**21**) are barely visible at all.

→ Lastly, on the perimeter of the lunar disc look at high magnification for the mountains that can be seen in profile.

1. Tycho	**8.** Manilius	**15.** Aristarchus
2. Copernicus	**9.** Bode	**16.** Posidonius
3. Kepler	**10.** Mösting A	**17.** Aristoteles/Eudoxus
4. Langrenus	**11.** Birt	**18.** Archimedes
5. Proclus	**12.** Pytheas	**19.** Gassendi
6. Dawes	**13.** Gambart A	**20.** Theophilus
7. Dionysius	**14.** Euclides	**21.** Ptolemaeus

From Full Moon to New Moon

To observe the Moon late in the evening, after Full Moon, you will have to wait until later for it to be high enough above the horizon.

At the end of the night the Moon will be higher and higher in the southern sky until Last Quarter. Then it will decline to the east between Last Quarter and New Moon.

Morning observation often benefits from minimum turbulence because the heat stored in the ground has been given up over the first part of the night much reducing local disturbance of the air. In addition, low-pressure areas apart, the weather is usually calmer in the second part of the night. So the conditions are right for better-quality images and it would be a pity not to make the most of it.

This is the ideal period for observing again all the features you have discovered over 14 consecutive nights, but this time in reverse order. The landforms are now lighted from the other direction, from east to west. And new details will appear in each formation, above all in the asymmetrical ones such as the rilles and scarps. But irregular craters and domes also display interesting changes. The differences between the east and west walls of certain large craters can be more readily appreciated.

Observations are made along the 'evening terminator' because these are the regions where the Sun is setting on the Moon.

MATCHING OBSERVATIONS

One day after Full Moon	→	Evening 1 (p. 36)
Two days after Full Moon	→	Evening 2 (p. 36)
Three days after Full Moon	→	Evening 3 (p. 38)
Four days after Full Moon	→	Evening 4 (p. 46)
Five days after Full Moon	→	Evening 5 (p. 54)
Six days after Full Moon	→	Evening 6 (p. 62)
Last Quarter	→	Evening 7 (p. 70)
One day after Last Quarter	→	Evening 8 (p. 84)
Two days after Last Quarter	→	Evening 9 (p. 94)
Three days after Last Quarter	→	Evening 10 (p. 102)
Four days after Last Quarter	→	Evening 11 (p. 110)
Five days after Last Quarter	→	Evening 12 (p. 118)
Six days after Last Quarter	→	Evening 13 (p. 128)

1. Mare Frigoris
2. Plato
3. Sinus Roris
4. Mare Imbrium
5. Archimedes, Aristillus and Autolycus
6. Apennine Mountains
7. Aristarchus
8. Eratosthenes
9. Copernicus
10. Kepler
11. Oceanus Procellarum
12. Grimaldi
13. Ptolemaeus, Alphonsus and Arzachel
14. Mare Nubium
15. Mare Humorum
16. Tycho
17. Schickard
18. Clavius

Further reading for Moon-watchers

Moon maps

• Orthogonal projection
Shows how features look from Earth. Ideal for observation.
- Association française d'astronomie Moon map (1:20 000 000 map – drawn ground and explanatory booklet).
- Phillips map (1:7 700 000 – line-drawn ground).
- Hallwag map (1:5 000 000 – photo ground).
- Lunar quadrant maps (4 × 1:3 500 000 – 1980 – on line-drawn ground).
- Rand McNally map (1:2 300 000 – 1980 – on photo ground).

• Graphical projection
Shows the true shape of features. Less useful for observation but educational.
- NASA Lunar Astronautical Charts (68× 1:1 000 000 maps - aerograph ground).
- IGN map (1:8 000 000 – aerograph ground).

Lunar atlases

- *Lunar Orbiter Photographic Atlas of the Moon*, D.E. Bowker and J.K. Hughes, NASA SP-206, US Government Printing Office, 1971.
- *The Hatfield Photographic Lunar Atlas,* J. Cook (ed.), Springer Verlag, 1998.
- *A New Photographic Atlas of the Moon*, Z. Kopal, Taplinger, 1971.
- *Photographic Lunar Atlas*, G.P. Kuiper *et al.,* University of Chicago Press, 1960.
- *Atlas of the Lunar Terminator*, John E. Westfall, Cambridge University Press, 2000.
- *Rectified Lunar Atlas: Supplement 2 to the Photographic Lunar Atlas*, E. A. Whitaker *et al.*, University of Arizona Press, 1963.

Books about the Moon

- *The Moon*, G. North, Cambridge University Press, 2000.
- *Epic Moon*, W.P. Sheehan and T.A. Dobbins, Willmann Bell, 2001.
- *Observing the Moon*, P.T. Wlasuk, Springer Verlag, 2000.
- *The Moon Book,* Kim Long, Johnson Books, 1988.
- *Observing the Moon with Binoculars and Small Telescopes,* E. Cherrington, Dover, 1994.
- *A Portfolio of Lunar Drawings,* Harold Hill, Cambridge University Press, 1991.
- *The Moon Observer's Handbook,* F.W. Price, Cambridge University Press, 1988.
- *The Moon: Our Sister Planet,* P. Cadogan, Cambridge University Press, 1981.
- *Patrick Moore on The Moon,* P. Moore, Cassell, 2001.
- *Who's Who on the Moon,* E. and J. Cock, Tudor Publishers, 1995.
- *Mapping and Naming the Moon,* E.A. Whitaker, Cambridge University Press, 1999.
- *The Motion of the Moon,* Alan H. Cook, Adam Hilger, 1988.
- *To a Rocky Moon,* D.E. Wilhelms, University of Arizona Press, 1993.

Software and CD-ROMs

Virtual Moon, Christian Legrand and Patrick Chevalley (freeware).
Clementine photos CD-ROM, NASA.
Lunarcalc, Alister Ling (freeware).

Internet sites (some of many)

- Moon Internet Sites list:
http://www.islandnet.com/~pjhughes/sol3.htm#moon
- Association of Lunar and Planetary Observers (ALPO):
http://www.lpl.arizona.edu/~rhill/alpo/lunar.html
- American Lunar Society:
http://otterdad.dynip.com/als/
- Lunar and Planetary Institute:
http//cass.jsc.nasa.gov/moon.html
- United States Geological Survey (USGS):
http://www.flag.wr.usgs.gov/usgsflag/space/wall/moon.html
- Lunar Orbiter IV Atlas of the Moon:
http://cass.jsc.nasa.gov/research/lunar_orbiter
- Lunar Exploration Timeline:
http://nssdc.gsfc.nasa.gov/planetary/chrono1.html

Glossary

achromatic : astronomical refractor objective optic made of two lenses of crown and flint glass that diminish but do not completely eliminate chromatic aberration.

apochromatic : astronomical refractor objective optic with two or three lenses of special glass free from chromatic aberration.

Dobsonian : type of Newtonian telescope on altazimuth mount, generally of large aperture, for deep-sky observation.

gibbous (Moon) : term designating the 'bulging' Moon between First Quarter and Full Moon and between Full Moon and Last Quarter.

two-star method : method of precisely aligning a telescope by successive observation of the movements of two very widely spaced stars.

orthoscopic : type of eyepiece with a set of triple lenses and a single lens.

Plössl : type of eyepiece with two sets of double lenses.

focal ratio : ratio of a telescope's focal length to its aperture, characterising its light grasp. The smaller the figure, the greater the telescope's light grasp.

terminator : the borderline between day and night on the surface of the Moon. The 'morning terminator' is the region where the sun is rising on the Moon. For an Earth-bound observer, it is observed in the evenings between New Moon and Full Moon. The 'evening terminator' is the boundary line where the Sun is setting and is observed from Earth in the mornings between Full Moon and New Moon.

turbulence : disturbance of the layers of the Earth's atmosphere causing the images of objects seen through a telescope to quiver.

Latin and English names

Aestatis (Lacus) : Lake of Summer
Aestuum (Sinus) : Bay of Heats
Asperitatis (Sinus) : Bay of Roughness
Australe (Mare) : Southern Sea
Cognitum (Mare) : Sea of Knowledge
Crisium (Mare) : Sea of Crises
Epidemiarum (Palus) : Marsh of Epidemics
Fecunditatis (Mare) : Sea of Fertility (or Plenty)
Frigoris (Mare) : Sea of Cold
Humboldtianum (Mare) : Humboldt's Sea
Humorum (Mare) : Sea of Humours (or Moisture)
Imbrium (Mare) : Sea of Showers (or Rains)
Insularum (Mare) : Sea of Islands
Iridum (Sinus) : Bay of Rainbows
Marginis (Mare) : Marginal Sea
Medii (Sinus) : Central Bay

Mortis (Lacus) : Lake of Death
Nectaris (Mare) : Sea of Nectar
Nubium (Mare) : Sea of Clouds
Orientale (Mare) : Eastern Sea
Procellarum (Oceanus) : Ocean of Storms
Putredinis (Palus) : Marsh of Decay
Roris (Sinus) : Bay of Dews
Serenitatis (Mare) : Sea of Serenity
Smythii (Mare) : Smyth's Sea
Somnii (Palus) : Marsh of Sleep
Somniorum (Lacus) : Lake of Dreamers
Spumans (Mare) : Foaming Sea
Tranquillitatis (Mare) : Sea of Tranquillity
Mare Undarum : Sea of Waves
Vaporum (Mare) : Sea of Vapours

Index

The bold figures refer to sites shown in the photos and described in the text. The others refer to sites shown in photos only.

Photo credits

Pp. 8–9, NASA / *Ciel & Espace*; p. 10, NASA / *Ciel & Espace*; p. 11, NASA / *Ciel & Espace*; p. 12, NASA / *Ciel & Espace*; p. 13, NASA / *Ciel & Espace*; p. 13 right, SIC / *Ciel & Espace*; pp. 18–19, A. Cirou / *Ciel & Espace*; p. 21, left and right, C. Legrand; p. 24, E. Graeff / *Ciel & Espace*; p. 26. A. Cirou / *Ciel & Espace*; p. 27, C. Legrand; p. 28, C. Legrand; p. 29, C. Legrand; p. 31, C. Legrand; p. 33, right, T. Legault; pp. 34–35, SIC / *Ciel & Espace*; pp. 36–37, E. Lecoq; pp. 38–39, E. Lecoq; pp. 40–41, C. Ichkanian; pp. 42–43, UCO / Lick Observatory image; pp. 44–45, E. Lecoq; pp. 46–47, E. Lecoq; pp. 48–49, E. Lecoq; pp. 50–51, E. Lecoq; pp. 52–53, C. Ichkanian; pp. 54–55, E. Lecoq; pp. 56–57, E. Lecoq; pp. 58–59, C. Ichkanian; pp. 60–61, C. Ichkanian; p. 62 top and p. 63, E. Lecoq; p. 62, below: NASA / *Ciel & Espace*; pp. 64–65, E. Lecoq; pp. 66–67, M. Jousset; pp. 68–69, C. Ichkanian; pp. 70–71, E. Lecoq; pp. 72–73, C. Ichanian; pp. 74–75, M. Jousset; p. 75 top, NASA / *Ciel & Espace*; pp. 76–77, M. Jousset; pp. 78–79, M. Jousset; pp. 80–81, C. Ichkanian; pp. 82–83, M. Jousset; pp. 84–85, E. Lecoq; pp. 86–87, M. Jousset; pp. 88–89, E. Lecoq; pp. 90–91, C. Ichkanian; pp. 92–93, E. Lecoq; pp. 94–95, E. Lecoq; pp. 96–97, M. Jousset; pp. 98–99, E. Lecoq; p. 99, NASA / *Ciel & Espace*; pp. 100–101, C. Ichkanian; pp. 102–103, UCO / Lick Observatory image; pp. 104–105, C. Ichkanian; pp. 106–107, E. Lecoq; pp. 108–109, E. Lecoq; pp. 110–111, E. Lecoq; pp. 112–113, C. Ichkanian; pp. 114–115, C. Ichkanian; pp. 116–117, E. Lecoq; pp. 118–119, E. Lecoq; pp. 120–121, C. Ichkanian; pp. 122–123, E. Lecoq; p. 123 top, Edimedia; pp. 124–125, E. Lecoq; pp. 126–127, UCO / Lick Observatory image; pp. 128–129, UCO / Lick Observatory image; pp. 130–131, E. Lecoq; pp. 132–133, UCO / Lick Observatory image; pp. 134–135, C. Ichkanian.

Flap maps of the Moon, UCO / Lick Observatory image.